# Conducting Polymers

## Special Applications

# Conducting Polymers

## Special Applications

*Proceedings of the Workshop
held at Sintra, Portugal, July 28–31, 1986*

Sponsored by the
U.S. Army Research, Development and Standardization Group
(U.K.)

Edited by

### LUIS ALCÁCER

*Chemistry Department,
Instituto Superior Técnico, Lisbon, Portugal*

## D. REIDEL PUBLISHING COMPANY

A MEMBER OF THE KLUWER  ACADEMIC PUBLISHERS GROUP

DORDRECHT / BOSTON / LANCASTER / TOKYO

**Library of Congress Cataloging in Publication Data**

Conducting polymers, special applications.

Includes index.
1. Organic conductors—Congresses.  I.  Alcácer, Luis, 1937–
QD382.C66C66      1987      620.1′9204297      87–9644
ISBN 90–277–2529–2

---

Published by D. Reidel Publishing Company,
P.O. Box 17, 3300 AA Dordrecht, Holland.

Sold and distributed in the U.S.A. and Canada
by Kluwer Academic Publishers,
101 Philip Drive, Assinippi Park, Norwell, MA 02061, U.S.A.

In all other countries, sold and distributed
by Kluwer Academic Publishers Group,
P.O. Box 322, 3300 AH Dordrecht, Holland.

# TABLE OF CONTENTS

LIST OF PARTICIPANTS ............................... vii

PREFACE ............................................. ix

J. O'M. BOCKRIS and DAVID MILLER / The Electrochemistry
        of Electronically Conducting Polymers .......... 1

S. LEFRANT / In Situ Raman Experiments on Polyacetylene
        in Electrochemical Cells ....................... 37

G. GEIB, H. KEYZER and K.G. REIMER / Conductive Comple-
        xes of Novel Porphyrin and Phenothiazine Poly-
        mer Systems .................................... 47

N. THEOPHILOU and H. NAARMANN / Increasing the Conductivity
        of Polyacetylene Films by Elongation ........... 65

P.D. TOWNSEND / Anisotropic Properties of Oriented Durham
        Route Polyacetylene ............................ 77

R. HUQ, L-L. YANG and G.C. FARRINGTON / Ionically-Conductive
        Solid Solutions of Divalent Cation Salts in Poly-
        (Ethylene Oxide) ............................... 89

J.R. STEVENS and B.E. MELLANDER / Room Temperature High
        Ionic Conductivity from Alkali Metal - Silver
        Halide - Poly(Ethylene Oxide) Complexes ........ 95

J. OWEN / Polymers with Both Ionic and Electronic Conducti-
        vity ........................................... 103

A.G. MACDIARMID, J.C. CHIANG, A.F. RICHTER, N.L.D. SOMASIRI
        and A.J. EPSTEIN / Polyaniline: Synthesis and
        Characterization of the Emeraldine Oxidation State
        by Elemental Analysis .......................... 105

ARTHUR J. EPSTEIN, JOHN M. GINDER, ALAN F. RICHTER and
        ALAN MACDIARMID / Are Semiconducting Polymers
        Polymeric Semiconductors ? : Polyaniline as an
        Example of "Conducting Polymers" ............... 121

F. WUDL / New Electronically Conducting Polymers ........ 141

SAMSON A. JENEKHE / New Electronically Conducting Polymers:
    Effects of Molecular Structure on Intrinsic Elec-
    tronic Properties ............................          149

R. SPINDLER and D.F. SHRIVER / High Conductivity in an
    Amorphous Crosslinked Siloxane Polymer Elec-
    trolyte .......................................         151

ROBERT J. MAMMONE / Use of Electronically Conducting Poly-
    mers as Catalytic Electrodes in Aqueous and
    Inorganic Electrolytes ........................         161

MICHAEL HANACK, SONJA DEGER, UWE KEPPELER, ARMIN LANGE,
    ANDREAS LEVERENZ and MANFRED REIN / Polyphta-
    locyanines ....................................         173

C.G. MORGAN, Y.P. YIANNI and S.S. SANDHU / Langmuir-
    Blodgett Deposition of Amphipathic Azobenzene
    Compounds for Surface Activation and Fabrica-
    tion of Functionalised Thin Films .............         179

J.J. MIASIK, A. HOOPER, P.T. MOSELEY and B.C. TOFIELD/
    Electronically Conducting Polymer Gas Sensors ...       189

REPORTS ON THE PANEL DISCUSSIONS .......................    199
    1 - Preparative Methods and Properties of Well Orient-
        ted Electronically Conducting Polymers ............  201

    2 - Prospectives of Realization of Polymer Electrolytes
        with Amorphous Structures and Consequently High
        Conductivity at Room Temperature .................  205

    3 - Interfaces and Contact Problems ..................  207

    4 - Cyclability of Polymer Electrolyte Cells. Power
        Efficiency and Energy Density ...................  209

CONCLUSIONS .............................................    211
    Conclusion on Electronically Conducting Polymers ......  213
    Conclusion on Ionically Conductive Polymers ...........  215

SUBJECT INDEX ..........................................     217

# LIST OF PARTICIPANTS

Teresa Abrantes, Technical University of Lisbon, Portugal
Iqbal Ahmad, U.S. Army Res. Dev. & Stand. Group, U.K.
Luis Alcácer, Technical University of Lisbon, Portugal
Vitor Amaral, University of Oporto, Portugal
Ray Baughman, Allied Signal Inc., U.S.A.
Veronica Bermudez, Technical University of Lisbon, Portugal
John Bockris, Texas A & M University, U.S.A
Arthur Epstein, Ohio State University, U.S.A.
Gregory Farrington, University of Pensilvania,Philadelphia, U.S.A.
Michel Gauthier, Hydro-Quebec, Canada
Olgun Güven, Hacettepe University, Turkey
Michael Hanack, University of Turbingen, West Germany
Rui Henriques, L.N.E.T.I., Sacavem, Portugal
Samson Jenekhe, Honeywell Inc., Bloomington Minn., U.S.A.
Hartmut Kahlert, University of Graz, Austria
Hendrik Keyzer, Cal. State University, U.S.A.
Serge Lefrant, University of Nantes, France
Alan MacDiarmid, University of Pensilvania, Philadelphia, U.S.A.
Robert Mamonne, U.S. Army E.T.D.L., Fort Monmouth, N.J., U.S.A.
Manuel Matos, L.N.E.T.I., Sacavem, Portugal
Jorge Morgado, Technical University of Lisbon, Portugal
Chris Morgan, University of Salford, U.K.
Herbert Naarmann, B.A.S.F., Ludwigshafen, West Germany
John Owen, University of Salford, U.K.
Marta Ramos, University of Braga, Portugal
Daniel Rueda, University of Madrid, Spain
Bruno Scrosati, University of Rome, Italy
Hideki Shirakawa, Univeristy of Tsukuba, Japan
Ralph Spindler, Northwestern University, Eavanston III, U.S.A.
Jim Stevens, University of Guelph, Ontario, Canada
Pat Moseley, A.E.R.E., Harwell, U.K.
Ana Stack, Technical University of Lisbon, Portugal
Fernando Teixeira, L.N.E.T.I. Sacavem, Portugal
Paul Townsend, Cavendish Laboratory, Cambridge, U.K.
David Venezky, Office of Naval Research, London
Fred Wudl, University of California, Santa Barbara, U.S.A.
Kenneth Wynne, Office of Naval Res. Washington, U.S.A.

# PREFACE

The development and the study of both ionic and electronically conducting polymers have been, in the past few years, areas of increasing interest. These new materials are, in fact, being considered for many technological applications, namely low weight, high energy density batteries and sensors.

This volume contains the proceedings of a workshop on this subject, sponsored by the U.S. Army Research, Development and Standardization Group (U.K.), which took place in Sintra - Portugal from July 27 to July 31, 1986.

The workshop, which included lectures, communications and discussion panels, was very sucessfull and the combination of ionic with electronically conducting polymers and their applications, not usually together in workshops or conferences, proved to be an excellent idea.

Lisbon December, 1986

Luis Alcacer

# THE ELECTROCHEMISTRY OF ELECTRONICALLY CONDUCTING POLYMERS

J. O'M. Bockris and David Miller
Department of Chemistry
Texas A&M University
College Station, Texas   77843   USA

ABSTRACT.   The new field of the electrochemistry of electronically
conducting polymers is reviewed.   A brief historical account traces
the beginning of organic electrodes to Kallmann and Pope, who, in 1960,
observed charge injection and conductance in anthracene electrodes.
The progress in the discovery and use of new polymer electrodes is
briefly discussed.   Some of the possible applications of these new
electrodes are suggested.   As important background information for
studying organic polymer electrochemistry, knowledge of the conduction
mechanism is needed.   The theory of bipolaron formation, as proposed
by Bredas, et al., is presented.   It is important to study the
electrode-solution interface.   Double layer models for metal, semiconductor, and insulator electrodes are probed.   Recent work and
applications of these electrodes are then briefly reviewed.   This
includes initiatives in the fields of electrode generated reactions,
photoelectrochemistry, batteries, and molecular electronics.   Finally,
the needed areas of research, from an electrochemical point of view,
are presented.

## 1.   INTRODUCTION

The phrase, electronically conducting polymers, awakens in the minds
of electrochemists such a large and overwhelming swathe of possible
advances that it is difficult to avoid a feeling of excitement in
considering it.   It is desirable to recognize those whom we conceive to
be the Grandfathers of this field and they lie further in the past than
is often thought (Fig. 1).   We believe it is reasonable to mention here
the names of Kallmann and Pope (1) who, in the Journal of Chemical
Physics of 1960, published a paper which for the first time showed the
passage of electric current across the interface between substances
which had been previously regarded as highly nonconducting, for
example anthracene, and the electrolytic solution.   This classical
paper stands at the base of attempts to encourage the use of organic
materials as conducting electrodes.   Kallmann and Pope put forward a
theory in which the resistivity of anthracene in the dry is radically
reduced by the injection of charge carriers from iodine in solution.

*L. Alcácer (ed.), Conducting Polymers, 1–36.*
© *1987 by D. Reidel Publishing Company.*

1960    Anthracene-Kallmann and Pope

1977    Polysulfur Nitride-MacDiarmid, Nowak, Mark, and Weber

1980    Polypyrrole-Diaz and Castillo

1980    Polyaniline-Diaz and Logan

1981    Polyacetylene-MacDiarmid, Nigrey, MacInnes and Nairns

1983    Polyphenylene-Shacklette, Elsenbaumer and Baughman

1983    Polythiophene-Diaz, Waltman and Bargon

1986    Polyisothianaphthene-Wudl

Figure 1.   Time line of the use of electronically conducting
organic polymers as electrodes.

Thus, a piece of anthracene in solution can act as an electrode and
exchange charges with ions in the surrounding solution.  Later in the
60's, Mehl and Hale (2) did further work on conduction in anthracene
electrodes.  Although the potential differences across them were about
100 volts, these authors confirmed the concept of Kallmann and Pope
that organic materials of very small intrinsic conductivity in the dry
could be woken up by charge injection to act as electrodes.

The second stage of this endeavor, some might regard it as the
first, was introduced by A. G. MacDiarmid, at the University of
Pennsylvania, who synthesized polysulfurnitride films (3) and used them
as electrodes (4).  Conductivities of the films were calculated to be
$3 \times 10^3 \ (\Omega \ cm)^{-1}$ (3).  More well known, however, are the extensive
works of MacDiarmid and his co-workers on the electrochemistry of
polyacetylene (5).  Thus, energy storage possibilities - as also those
of fuel cells and photoconverters - were investigated at this institu-
tion during the last 10 years.

Of the several areas which beckon us towards the study of
electronically conducting polymers is the biological side, because in
this, one sees the potential of replicating body processes to a higher
degree than would be possible were it necessary to use metallic conductors in prosthetics.  Indeed, there appears to be a continuity between
the thinking of Szent-Gyorgyi (6) in the electronic conductance of
biological compounds (7) and that of Kallmann and Pope on electronically
conducting organics.

A large number of disciplines are involved in these studies, not
only physics and chemistry, materials science and biology, but also
particularly surface science studies and the studies of electron transfer at interfaces with the surrounding ionically conducting phase, i.e.,
the surface electrochemical phenomena (8).

My task today is to give an electrochemist's view of these developments.

## 2.   HISTORY

We have mentioned Kallmann and Pope (1) in the introductory remarks but it is easy to understate the importance of their 1960 paper which showed for the first time that shining light upon anthracene in contact with a solution containing iodide caused a change in current across the surface. This change was associated both with a change of conductivity of the solid and with a change in the electron transfer rate across the interface.

Mehl and Hale (2,9) concluded that electrochemical reactions would take place at the anthracene-aqueous solution interface so long as the compounds which are present in solution were able to inject holes into the valence band or receive electrons from the conduction band.

One of the descendants of the Mehl and Hale work was Boguslavski in the U.S.S.R. His paper (10) is remarkable in that he showed that even with Teflon, one of the more resistive materials known, there could be charge injection, and the corresponding use of the material as an electrode.

Thus, in Fig. 2, we see a current-voltage curve where Boguslavski has used Teflon as the electrode. Although the potentials amount to several volts, the measurement of significant current densities over 3 orders of magnitude is a remarkable achievement.

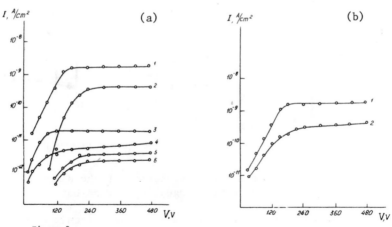

Figure 2
Dependence of the cathode current density on the applied voltage
(a) Teflon in various solutions: 1) $2.5 \times 10^{-2}$ M Ce$^{+4}$ in 1 M H$_2$SO$_4$;
2) $10^{-1}$ M KMnO$_4$ in 1 M H$_2$SO$_4$; 3) 1M H$_2$SO$_4$ + O$_2$; 4) 0.5 M H$_2$SO$_4$ + O$_2$;
5) 1 M NaCl + O$_2$; 6) 1 M NaOH + O$_2$.
(b) Different thicknesses of Teflon in $2.5 \times 10^{-2}$ M Ce$^{+4}$ in 1 M H$_2$SO$_4$:
1) 5 μm; 2) 40 μm.

Robert Nowak (4), associated with A. G. MacDiarmid and Allen Heeger, was the first in 1977 to use an electronically conducting polymer in electrochemical studies. Thus, the substance used was the

polymer of sulfur nitride and they were able to record cyclic voltammograms in a solution of 0.1 M $KNO_3$. They found substantially different
results in the voltammograms, depending on the parallel and perpendicular
position of the strands of the polymer.

Next was the work of Diaz (11), who in 1980 showed that polypyrrole
could act as an electrode. He found that the material could be turned
from a semiconductor to a conductor by changes of electrode potential,
that it was stable and non-porous* and that it was possible to obtain
voltammograms with it. A similar 1980 study of Diaz (13) is on
polyaniline, where cyclic voltammograms were measured in 0.1 $Et_4NBF_4$ in
acetonitrile in the presence of sodium chloride.

The work which has caught the attention of the commercial audience
is that of MacDiarmid, who in 1981 published a paper on the use of
polyacetylene in propylene carbonate (5). Thus, for the first time, a
battery involving polyacetylene (Fig. 3) was introduced. It was found
possible to make the polyacetylene either in semiconductor or metallic
form in respect to its conductivity, and that this depended on the doping by lithium perchlorate. The maximum conductivity was of the order
of $10^3$ mhos $cm^{-1}$ and the "doped" polymer was written in the form
$((CH)^{+y} (ClO_4)^{-}_y)_x$.

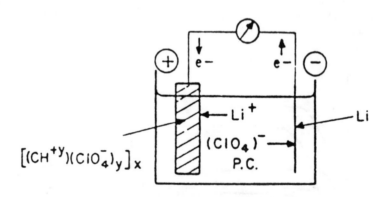

Figure 3. Schematic representation of the discharge process
in a $(CH)_x/LiClO_4/Li$ rechargeable storage battery cell.

They found that the battery had a high open circuit voltage of 3.7
volts and gave an initial current of 25 mA $cm^{-2}$. The overall energy
density was 176 watt hours per kilogram, compared with a value for the
lead acid battery of about 30.

More recently, Diaz (14) has studied polythiophene. He finds it
possible to work with a free standing film of polythiophene, or with

---

*More recent work (12) has shown some ionic permeability.

polythiophene on platinum. The doping in this case was carried out with a counter ion of $BF_4^-$ and variation of this greatly affected the electrical conductivity. The conductivity of polythiophene depends upon the oxidation state. This corresponds to removal of electrons from the polymer, and in the fully oxidized state one electron is removed per 4 thiophene units. Insofar as the polymer becomes charged positively during this oxidation, counter ions are necessary, and in the system used by Diaz these consist of $BF_4^-$.

The work of Diaz of 1983 is a reasonable point at which to end a description of the initial work in this field and turn to the modern phase.

## 3. BASIS OF THE PRESENT INTEREST IN THE ELECTROCHEMISTRY OF CONDUCTING ORGANIC POLYMERS

### 3.1 Electrodes

The overall interest in these materials arises with the background of the electrochemistry of metals where it can be taken for granted that the electronic conductivity will be large. However, there are some intrinsic disadvantages when dealing with metals in electrochemistry. Thus, as far as industrial applications are concerned, only iron and aluminum are cheap and abundant but both of these give rise to complex electrochemical interfaces, e.g., labile oxide formation. The more stable metallic electrodes are noble metals, where the cost is limiting.

Were it possible to develop the promise of electronically conducting organic polymers, an ehnanced horizon would be revealed. Thus, even in the face of oil and natural gas shortages, $CO_2$ could be used as a feedstock by reducing it electrochemically to useful compounds. The prospect for a long term supply of inexpensive, lightweight, non-corrodible electrode materials is hence improved.

### 3.2 Bioelectrochemistry

The possibility of an internally powered artificial heart would be enhanced by the availability of conducting polymers. Such a heart (15) would involve a pump which would be powered by a fuel cell using glucose from blood charged with products from nutrition as the fuel, together with an oxygen cathode in blood from the lungs (about 20 watts is needed from such a device).

One may foresee the possibility of other devices further along the research path: For example, organic conducting polymers may make it possible to make artificial nerves. Perhaps even modifications to the brain might eventually be contemplated.

### 3.3 Space and Air Flight

In space (and SDI) applications, light weight is at a premium. In respect to electrochemical and energy conversion (including photo-voltaic) devices, use of polymers with an average density of 1, rather

than from metals which have an average density nearer to 10, would be
attractive.

The power-to-weight- ratio of the internal combustion engine is
about 2 hp per pound, but the battery-electric motor combination is
about 10 pounds per hp, i.e., 20 pounds less advantageous, so a drop by
an order of magnitude of weight, probably attainable in electronically
conducting polymers, could give rise to ratios similar to an internal
combustion engine.

One might think also of a solar-driven aircraft: The availability
of photoactive organic polymers may give rise to photovoltaic "sails"
for the collection of solar energy in flight.

## 3.4  Military

3.4.1  Lightweight cells for long distance flights and for use in very
lightweight aircraft.

3.4.2  Portable arrays which could give energy conversion (stored in the
corresponding polymer batteries) in camps far from fuel supplies.

3.4.3  Use in devices where infrared detectability is a problem, e.g.,
tanks in hiding, etc.

## 4.   SOURCE OF ELECTRONIC CONDUCTION IN POLYMERS

The origin of the conductivity of polymers arises from a state of
relative oxidation or reduction (16,17).  In such states the polymer
itself loses (for oxidation) or gains (for reduction) electrons in its
structure.  The number of monomer units which gain or lose an electron
is variable but may be, e.g., 1 unit in 4.

Once the polymer is electronically charged, counter ions from
solution enter the polymer fibrils to produce electrostatic neutrality.
It is these ions which are often referred to as dopants.  However, this
is not doping in the sense of semiconductor doping, where the dopant
provides charge carriers.  In conducting polymers the charge carriers
are generated within the polymer chain.  On the other hand, it is convenient to refer to the counter ions in the charged polymers as dopants
so this term is widely used.

Considering the polymer in terms of a semiconducting material and
using the band structure model (18,19), oxidation and electron loss
gives rise to new energy states.  Removal of one electron from a π bond
leaves the remaining electron in a non-bonding orbital - different in
energy from the valence and conduction states.  These states are above
the valence band and they give rise to the behavior of the polymer as
though it were a heavily doped semiconductor.

Why was the existence of extensive semiconduction and even metallic
conduction in electronic polymers not discovered until the 1970's?  The
answer to this may be that it was not sought until then.  The first
discovery of high electron flow in polymeric materials was that of Labes
who noted it in polysulfurnitride (20).  MacDiarmid and Heeger's doping

of polyacetylene (5a) produced the first conducting <u>organic</u> polymer. This sparked a great deal of interest in this new field.

The relationship between conductivity and doping is switch like (21), as shown in Fig. 4.  At a certain degree of doping (i.e., a certain degree of unpaired electrons or oxidation) an avalanche-like increase of conduction occurs.

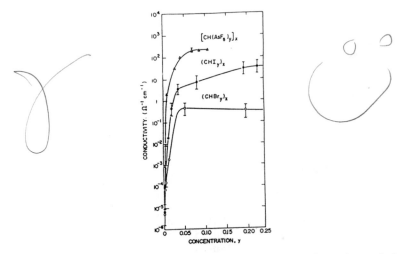

Figure 4.  Electrical conductivity as a function of dopant concentration.

The most commonly accepted model (16-19, 22-25) is that of polaron formation upon oxidation and combination of polarons into bipolarons. The bipolarons are then free to move along the polymer chain, which gives rise to the electronic conductivity.

Oxidation of the polymer breaks one double bond (Fig. 5), leaving a radical and a positive charge on the polymer chain, this is the polaron.  Polarons converge into bipolarons when the polaron concentration gets high enough for the polarons to "feel" each other, i.e., the radical-cations, spread out through adjacent $\pi$ structure across ∿ 8 bond lengths, makes contact with another radical cation.  The combination of the two radicals (one from each polaron), forms a new $\pi$ bond.  This $\pi$ bond is more stable than the two radical-cation bonds (i.e., the $\Delta G$ of the $\pi$ bond is greater than $\Delta G$ of dissociations of the two polarons). The result is a bipolaron which is more stable than two polarons at the same distance apart.

At low oxidation levels, coulombic repulsion of the positively charged polarons prevents the combination of the radicals which would lead to a bipolaron.  As the oxidation level is increased (caused by the externally applied potential difference) the concentration of polarons goes up and they become crowded together close enough for bipolaron formation to occur.  It is at this point in the oxidation process that the conductivity undergoes a marked increase.  Once the radical components of the polarons have combined to form $\pi$ bonds, the remaining positive charges achieve high mobility along the chain.

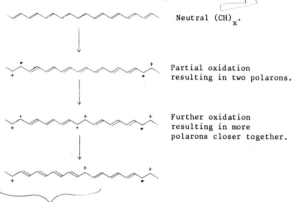

Neutral $(CH)_x$.

Partial oxidation resulting in two polarons.

Further oxidation resulting in more polarons closer together.

Close proximity of polarons results in interaction of two umpaired electrons to form a new pi bond and a mobile "bipolaron". Formation of the pi bond provides a "driving force".

Figure 5.   Bipolaron formation in polyacetylene.

However, it must be admitted that at this time it is not yet possible to make predictions of conductivities in organic metals and semiconductors.  For example, perylene $(AsF_2 . CH_2Cl_2)$ has a room temperature conductivity (26) of 2000 mho $cm^{-1}$, which is 4 times that of TTF TCNQ, up to that time the most conducting compound.  This degree of conduction could not have been predicted by the present theory.  Among the unsolved problems of the field one may cite the case of $(SN)_x$, which is a molecular substance, unionized, and highly conducting.  This contrasts to the fact that in most cases electronic polymers have to be made conducting by oxidation and ionization.  The field of neutral organic metals is limited to the above example, though the reasons for this are at present not known.  Research into very small band gap organic polymers, e.g., polyisothianaphthene (27), is continuing.

5.   LIKELY FUTURE APPLICATIONS OF ELECTRONICALLY CONDUCTING POLYMERS IN ELECTROCHEMICAL SCIENCE

5.1  Electronically Conducting Polymers As Electrodes

It is of interest to name here certain requirements which are necessary if polymers are to act as electrodes.

5.1.1  The requirement of conductivity is not a great one, for the polymer is generally put in thin layers onto the surface of a metallic conductor.  The potential difference across a layer of conductivity $\kappa$ is given, for a current density, i, by the equation:

$$\eta_{ohmic} = \frac{iL}{\kappa}$$

where "L" is the thickness of the layer and $\kappa$ is specific conductivity. Thus, if this potential difference is not to be more than 10 mv and the layer thickness is $10^{-4}$ cm then $\kappa$ can be as low as $10^{-4}$ mhos $cm^{-1}$.

5.1.2  Porosity:  If the polymer has as substrate a metal, the porosity must be zero.  The absence of porosity in polymer electrodes is diffi-cult to achieve.

If the polymer is too porous the electron transfer can take place at the substrate rather than at the polymer.  Bull et al. (28) found that this can be the situation when polypyrrole is used as the electrode.

If, on the other hand, the electrode attains sufficient mechanical stability, so that it may be used without a supporting substrate, then the porosity of the electrode is an advantage rather than a disadvantage for the affected electrode area will now increase and the polymer may become analogous to a porous electrode, similar to those used in fuel cells.  The·effective current density could be increased by about 2 orders of magnitude.

5.1.3  Surface area:  It is advantageous if the surface area can be high.  However, it must be high in the correct way.  For example, the usable area of deep pores may be limited firstly because they may allow contact of the electrolyte with the underlying substrate, and secondly because current lines in an electrode do not penetrate completely down the pores (8) and the length of the pores which remain active is limited, a subject well known in fuel cell electrochemistry.  SEM's of polymer surfaces are shown in Fig. 6.

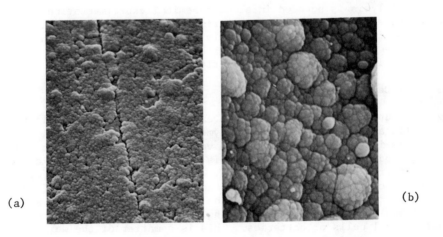

(a)                                                              (b)

Figure 6.   (a) SEM of polypyrrole/BF4 film, 10,000x, and (b) SEM of polythiophene/BF4 film, 15,000x.

5.1.4  Ohmic contacts:  Electronically conducting polymers as electrodes may be used with or without substrates.  If they are used

with substrates, it is important to make sure that the contact between
the polymer and the substrate is ohmic, i.e., that there is no exponen-
tial control of current arising from the contact between the substrate
and the polymer.  In the case of non-ohmic contacts, the properties of
the observed current–potential relationship will depend not only upon
the polymer–solution contact, but also upon the polymer–substrate contact.

5.1.5  A minimization of surface trap states:  In semiconductor
electrodes such as gallium phosphide, the control of the current at
the interface is influenced by the surface traps which exist there,
i.e., surface states which are outside the band gap.  In electronically
conducting polymers, surface traps inhibit the electrode reaction itself, and therefore must be minimized.

5.1.6  The presence of "good" surface states:  On the other hand,
surface states at semiconductor–solution interfaces may be "good" if
they exist within the band gap.  In electronically conducting polymers,
surface states which exist in the band gap ought to be encouraged, and
indeed they may be artificially introduced.
       It is worthwhile to draw attention here to the relationship between
electronically conducting polymers as electrodes and the idea that
enzymes can be regarded as electrodes, too (29).  This field has been
fully discussed by Tarasevich (30).

5.2  Electrocatalysis

Electrocatalysis is bounded in electrochemical experimentation by a
lack of materials with which to experiment.  The addition to these
materials of a large variety of easily adjustable substances in which
the surface groups could be varied almost infinitely in extent would
add a new dimension to the possibilities of experimentation.

5.3  Fuel Cells and Batteries

A gain of up to 1 order of magnitude can be expected in the weight of
fuel cells and batteries made of polymers.
       It is intended that sub–orbital vehicles will provide high speed,
high altitude flight.  A difficulty is the formation of $NO_x$ which could
be formed from burning hydrogen fuel.  The problem of the $NO_x$ formation
is associated with an attack on the ozone layer, the removal of which
would give rise to higher UV radiation incident on the earth's surface.
       It would therefore be interesting if it were possible to run air-
craft on fuel cells or batteries.  This is a matter of the weight and
power densities of the substances concerned.

5.4  Electrochemical Reactors

Were it possible to make highly diverse electrocatalysts which had
specificity introduced, i.e., introduction of artificial enzymes or the
active parts of enzymes, the possibility of electrochemical reactors, in

which the enzyme character of the electrodes could be easily changed, could give rise to a great number of applications.

## 5.5 Applications to Biochemistry and Biomedicine

One of the difficulties of the present situation in prosthetics is that metallic compounds used at present cause blood clotting. This clotting depends upon the degree and size of charge on the metal, and it has been shown by Sawyer and Srinivasan (31) that if the metal has a potential on the hydrogen scale of more negative than -.6V, blood does not clot on that metal and the prosthetic remains stable.

It seems likely that, given the great variety of surfaces which can be introduced into polymers, a greater array of possibilities for achieving a suitable surface charge exists if biopolymers are the basis for prosthetics rather than metals.

The concepts of an artificial heart can be looked at in terms of a fuel cell driving an electric motor-driven pump (32). This was a program supported by the NIH in the 1970's. The concept (Fig. 7) was that the heart could be run on a fuel cell consisting of one glucose-based electrode for oxidative power, the other electrode being used for the reduction of oxygen. A fuel cell was implanted in a dog, and ran for several days.

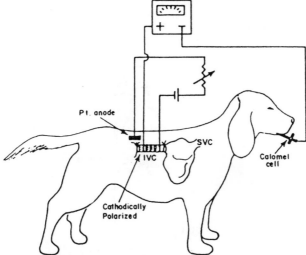

Figure 7. Biological Fuel Cell. Experimental arrangement for maintaining a copper prosthesis, implanted in the canine thoracic inferior vena cava.

## 5.7 Biosensors

The use of small chemical sensors that could be implanted in the body would be useful for analysis of conceivably any chemical in the body (33). This could provide analysis and a feedback mechanism (Fig. 8) for control of diabetes, etc. The use of electrodes as biosensors has

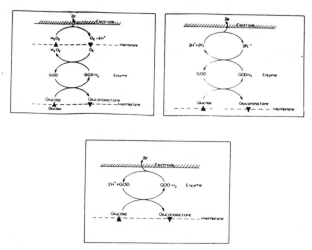

Figure 8.   Schematic of recent versions of biosensor operation.

been receiving attention (34).   The electrode can be given enzyme
properties (30,35) to promote electrocatalysis.   It is conceivable that
conducting organic polymers would be very useful as the biosensor (Fig.
9).   An organic polymer is more likely to interact with enzymes and
biochemicals compared to metal electrodes.

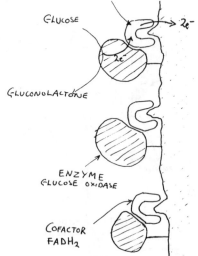

Figure 9.   Glucose sensor where the enzyme (containing a
reduced cofactor) is oxidized directly at an electrode.

## 5.7   Macro-reduction of Carbon Dioxide

The concept of a massive reduction of carbon dioxide as a source of
organic chemicals in the future when fossil fuels become economically

unattractive is one of interest. The only sufficiently abundant electrode material would be electronically conducting polymers, possibly synthesized from $CO_2$.

## 6. THE DOUBLE LAYER AT THE ELECTRONICALLY CONDUCTING POLYMER-SOLUTION INTERFACE

The interface at a metal-solution contact (36) is given by the diagram shown in Fig. 10. The potential difference at the interface exists between the sites on the metal surface and ions in the solution. This situation is generally considered to give rise to a "condensor-like-layer" which consists of two parts:

1.  The packed layer of charges within the metal itself, and
2.  The ions in solution which can be distinguished as those "in the Helmholtz layer" and a diffusely distributed layer of ions (the Gouy layer) out into the solution.

On the other hand, the semiconductor-solution interface has part of its double layer inside the substance concerned (Fig. 11). There are two versions of this semiconductor double layer. The more usually represented one shows a small potential drop from the electrolyte with most of the potential difference occurring in the charge cloud inside the electrode arising from the distribution of the positive and negative charges inside the semiconductor. A particular feature of this type of double layer is called "band bending"; the situation which exists with the energy levels of the conduction band and the valence band as the charge on the electrode is changed (37). The potential difference is supposed to change largely inside the semiconductor when the outside potential changes (Fig. 12).

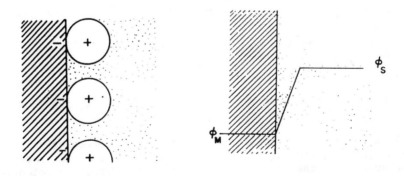

Figure 10.  Potential profile corresponding to the Helmholtz model.

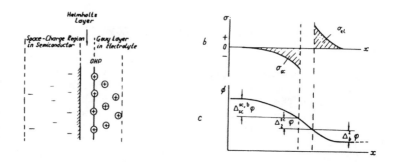

Figure 11. For a semiconductor electrode: (a) Double layer structure; (b) charge distribution; (c) potential drop.

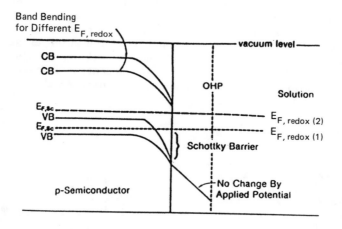

Figure 12. Schottky barrier model at the semiconductor inter-face.

This is the conventional picture of the semiconductor–solution interface but a different situation develops when surface states exist on the surface of the semiconductor (Fig. 13). Here, the potential difference which arises across the interface as the outside source sup-plies charges now goes largely on the Helmholtz layer, as with metals, and only to a small extent on the inside of the semiconductor.

The question which has excited semiconductor chemists in recent times is whether or not sufficient states exist ($10^{13}$ per square centi-meter) to make the approximation used in the latter model valid (see

below).   These two models can be distinguished by means of capacitance
measurements.

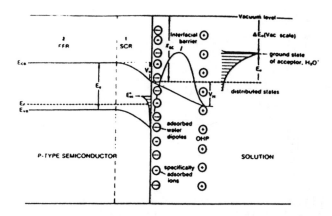

Figure 13.   Location of surface states in a p-type semiconductor.

## 6.1   Insulator-Solution Interfaces

The insulator-solution interface can be regarded as similar to that of a
semiconductor-solution interface except that, when the number of charge
carriers is small, corresponding to an insulator of say, $10^{10}$ charges
per cc, the space charge region becomes a large one, say, 1/kappa might
be in the region of $10^6$A.

The same question is involved with the insulator electrodes as
with the semiconductor electrodes.   Are there surface states and is the
model better represented with the charges largely inside the insulator
(Fig. 14) or do they settle predominantly on the surface?

## 6.2   The Determination of Surface States

Surface states can be determined for semiconductor-solution interfaces
(38) by making measurements of the capacitance of the semiconductor-
solution interface and then by producing equivalent circuits which
simulate the measurement to the best extent.   An example is given (Fig.
15) which shows the surface states for gallium phosphide as a function
of potential.   Here, concentration of surface states is large, corresponding to the second model of the interface.   This seems likely to be
the most general case because of the presence in the double layer of
adsorbed ion and solvent molecules.

The second kind of study which is helpful in understanding the
polymer-solution interface lies in the determination of adsorption.
This can be made in numerous ways.   The most recent, and perhaps the
most fruitful, is to use FTIR spectroscopy for in-situ detection of

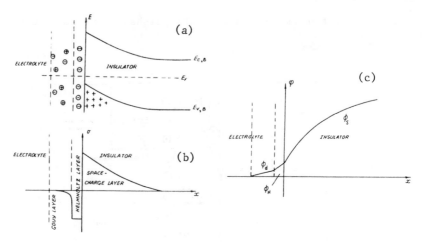

Figure 14. For an insulator electrode: (a) Band bending, (b) Charge distribution, (c) Potential drop.

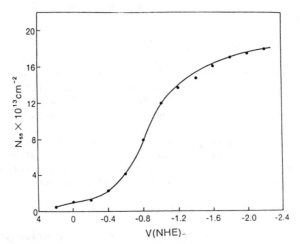

Figure 15. Surface state density as a function of bias potential for GaP-electrolyte interface.

adsorbed species. An example (39) of this method's capabilities is shown in Fig. 16. Sometimes this data can be analyzed in a form which allows the evaluation of the Gibbs-surface excess of anions, cations and neutral molecules since capacitance is a function of the ion concentration in solution.

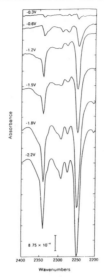

Figure 16.   FTIR determination of $CO_2$ and $CH_3CN$ adsorption on Pt.


## 7.   ELECTROCHEMICAL KINETIC WORK ON ELECTRONICALLY CONDUCTING POLYMERS

The general situation at this time is that the field of electrode-kinetic measurements is in a state of infancy.  One of the difficulties is illustrated by the work of Bull et al., who found (28) that polypyrrole-coated platinum appeared to be active because of the porosity of the polypyrrole film – the underlying substrate was in fact the active component.  Correspondingly, Feldberg found (40) that poly-pyrrole on Pt is porous to solvent which results in an unexpectedly large surface area.  From cyclic voltammograms he concludes (41) that significant proportions of the currents were taken up capacitively as the sweep rate increased.

The complexities of dealing with polyacetylene in solution was shown by Dietz and Beck (42).  The first peak in their voltammograms, e.g., in perchloric acid, showed a relatively small electron transfer (0.04 electrons per mole of CH).  A second peak showed the degradative oxidation of the polymer.

Audebert and Bidan (43) introduced halogen substitution into polypyrrole.  This gave rise to greater possibilities of introducing charge onto the polymer as shown by the higher doping.  The authors interpret the observed phenomenon as due to the inductive effect of the halogens and more resonance structures (Fig. 17).

Barendrecht (44) observed the oxidation and reduction of hydro-quinone at polypyrrole electrodes (Fig. 18).  Electrocatalytic effects are observed.  Although the quinone-hydroquinone couple is irreversible on gold, it is more reversible on the polypyrrole covered gold.  Cor-respondingly, molecular oxygen reduction has been observed by Barendrecht (45) on polypyrrole electrodes but here the reduction oc-curred largely on the metal because the polypyrrole was permeable.

Because of the increase in the mean residence time of the peroxide
near the metal/polypyrrole interface, the rate of reduction of oxygen
to water is greater than at the uncovered metal.

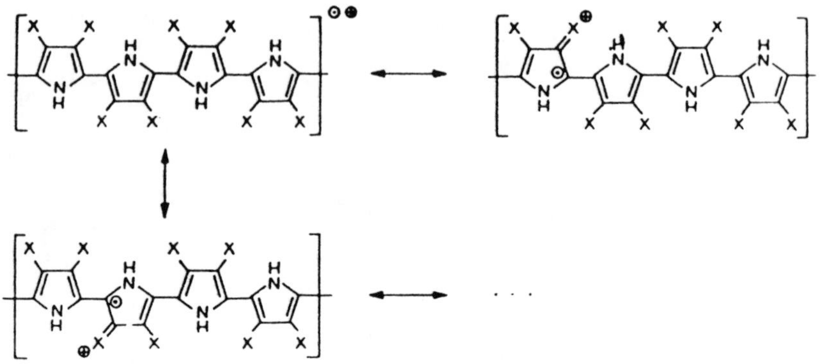

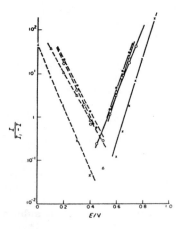

Figure 17.   Possible resonance structures for halogenated
polypyrrole.

Figure 18. $I/(I_1-I)$ vs. E for $H_2Q$ oxidation (——) and Q reduction (----) in
$H_2SO_4$/ethanol.   Uncovered Au disk (x); PP(Au) electrodes with
formation charges: 0.30 kCm$^{-2}$ (o), 0.60 kC m$^{-2}$ (Δ) and 1.20 kC m$^{-2}$(•).

Work has been carried out by Salmon, et al. (46).  They have looked
into redox reactions involving ferrocene largely with the idea of
evaluating substituent effects.  They determined the electrode substituent effect on the ferrocene couple reaction and thought that the
varying results were due to structure variations in the monomer.

Ewing et al. did work (47) showing oxidation of ascorbic acid and dopamine occur at the polypyrrole-electrolyte interface. They discuss possible electrostatic interactions between the anionic solutes in polypyrrole and the negative functional groups of the substrates.

## 8.    PHOTOELECTROCHEMICAL EXPERIMENTS USING ELECTRONICALLY CONDUCTING POLYMERS

One of the aims of the energy oriented electrochemistry laboratory (48) is to decompose water by means of light. The major difficulty is that the materials which are used at the present time are too expensive. Thus, the use of electronically conducting polymers is of great interest.

The first paper (49) in this area was by MacDiarmid and coworkers. They used polyacetylene in a photoelectrochemical cell. Illumination of the electrode produced a current of 40 $\mu$A/cm$^2$. The efficiency was about 3%.

Similar results were reported (50) by Shirakawa et al., utilizing a cell as shown in Fig. 19. Their results for dark and light illumination are shown in Fig. 20 and their current densities of about 9 microamps cm$^{-2}$. Here the conversion efficiency was better than that reported by MacDiarmid and appears to be about 6%.

Thus the Japanese work seems to offer possibilities of light initiation of reactions which would be of the same order as those which are being obtained now with much more expensive inorganic materials.

## 9.    THE USE OF ELECTRONICALLY CONDUCTING POLYMERS AS BATTERY ELECTRODES

The first paper in the field was indeed by MacDiarmid (51) who described in 1981 a battery consisting of polyacetylene "doped" with perchlorate, and having a counterelectrode made of lithium.

The equation

$$[CH^{+0.06}(ClO_4)^-_{0.06}]_x + 0.06xLi \rightarrow (CH)_x + 0.06xLiClO_4$$

These workers reported an open circuit potential with a remarkably high value of 3.7 volts. The value of the current which they withdrew from the cell did not seem to be well established. An extraordinarily high power density (680 W/kg) is also stated to have been "estimated" but the cell potential decay from 3.7V to 2.7V over 25 minutes while the discharge current is held at 0.5 mAcm$^{-2}$.

This work was reviewed by Somoano (52), who pointed out that one of the limitations on the depth of discharge which one can take out of the cell is the transition from a metal to a semiconductor that occurs at some point during the discharge. Thus, the depth of discharge would be limited due to the minimum value of oxidation that is needed to retain high conductivity.

The cell attributed to MacDiarmid above was turned into a cell with opposite characteristics in the paper (53) of 1982. Here the

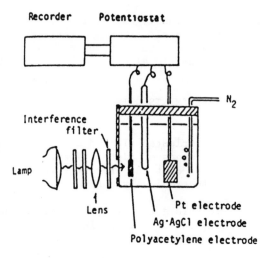

Figure 19. Experimental apparatus used by Shirakawa for photoelectrochemical measurements of a polyacetylene electrode.

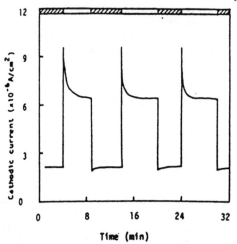

Figure 20. Cathodic current derived from a trans polyacetylene electrode in the dark and on illumination.

polyacetylene was doped in an opposite way to that described earlier (i.e., n-type). Lithium ions (instead of the corresponding perchlorate ion) were introduced into the polyacetylene structure so that the polyacetylene itself can be written as

$$[Li_y^+(CH^{-y})]_x$$

In this case the polyacetylene starts out in the neutral state. Spontaneous reduction of polyacetylene occurs as the lithium dissolves and dopes the polyacetylene to give the above formula. The electrolyte is lithium perchlorate in THF.

A number of other battery types have been published (54) by MacDiarmid, for example, a dual polyacetylene electrode battery has been described in which there is a $p(CH)_x$ cathode and an $n-(CH)_x$ anode in 1 M $LiClO_4$ with the solvent being propylene carbonate or sulfolane. Here the individual reactions are

$$(CH^{+0.06})_x + 0.06 \text{ e} \rightarrow (CH)_x$$

$$(CH^{-0.06})_x \rightarrow (CH)_x + x0.06 \text{ e}.$$

A good short circuit current of 100 mA cm$^{-2}$ is found only for a circuit potential of 2.4V. Shelf life was short. The current of cells similar to those shown decline to about half the original value in a few hours (Fig. 21).

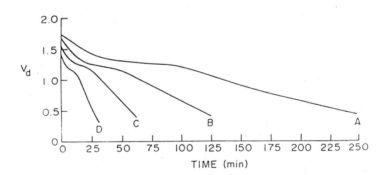

Figure 21. Constant current discharge characteristics of polyacetylene reduced to 10.0 mol%, i.e., $[Li^+_{0.1} (CH)^{-0.1}]_x$ at various currents. A) 0.16mA B) 0.32mA C) 1.28mA.

Work similar to that done at the University of Pennsylvania was taken up by Diaz and his team (55) using polyazulene and polythiophene as new electronically conducting polymers. Open circuit potentials for polythiophene, 0.5V, are not as good.

Fritz Will has made a study of discharge behavior of polyacetylene electrodes (56) and has drawn attention to the fact that slow dopant diffusion and slow discharge at the polymer/electrolyte interface is the principal problem. No steady state potentials were obtained over about 1 hour. This work quantitates some of the difficulties with these batteries.

A recent paper of MacDiarmid involves polyaniline (57) and here the counter electrode is zinc. 100 mA per cm$^{-2}$ is the current density

and the open circuit voltage is 1.4.  A life of about 6 hours is
claimed.  This battery does utilize an aqueous electrolyte.

Kitani has also reported (58) a study of aqueous polyaniline
batteries using zinc as a counter electrode.  The corresponding values
of the energy density for the Japanese work was about 140 W hr/kg and
the current at 1 mA cm$^{-2}$.  The Japanese workers also used lead oxide
as a counter electrode and reported "excellent stability".  But the
maximum time shown in the paper is only 60 minutes.

A somewhat different standard of work has been described recently
in the paper (59) of Mermilliod et al., from the labs at Saclay in
France.  They have studied chemically synthesized pyrrole as electrode
materials from the point of view of battery manufacture and characterized them with cyclic voltammetry, AC impedance measurements and
chronopotentiometry.  Instead of regarding the devices as batteries
involving two electrodes, each with charge transfer reactions taking
place, the French workers introduced the idea that a capacitance forms
a considerable part of the energy storage properties of the cell.  The
capacitance rises from about 0.01 F to about 0.2 during a change of
potential of about 1 volt (Fig. 22).  They also find some correlation
between the redox potential and the overall capacity of the battery,
showing the redox potential as high when the capacity is low.

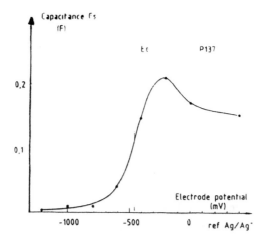

Figure 22.  Variation of the low frequency capacitance $C_s$ vs.
the potential of the polypyrrole electrode.

The French work gives an important beginning to a more detailed
analysis of the polypyrrole electrode and attempts to differentiate the
part played by the electrochemical reactions and that by the charge
storage.  They think that the capacitance represents about 25% of
the total electrode energy storage.

It is clear that the main things missing in this field are:
1.    An understanding of the kinetics in terms of discharge and
      diffusion.

2.  A quantitation of the capacitative and charge transfer effects involved in charge storage. This work should include investigation into increasing the usable potential range of the polymer electrodes and increasing the rate of charge transfer in order to promote higher battery currents.

## 10.  "MOLECULAR ELECTRONIC" DEVICES

There has been much discussion during the last 5 to 6 years of the possibility of miniaturizing electronic devices by decreasing component size to the molecular level, and major conferences have taken place (60) to discuss these matters.

The electronic circuits themselves would be composed of molecules that would individually act as wires, diodes, transistors, and memory devices. Conducting polymers would likely be used (60a) in many of these components. The polymers can be modified to give them memory states which can act by change of conformation or charge flow. A memory device could be addressed by current flow through the polymer, stimulated by an externally applied electric field, or by laser signal, etc.

An early example of an electrochemical rectifier based on polymers was reported by Lovrecek, Despic and Bockris (61) in 1959. The rectifier used polyvinyltoluenesulfonic acid and polyvinyltrimethylbenzylammonium hydroxide to allow ion flow in one direction only.

Some recent advances in this field have been due to Wrighton, who showed in 1984 a polypyrrole based transistor (62). The basic idea in Wrighton's work is to switch on a large current through the polymer by controlling the oxidation of the polymer. Polypyrrole was used on three gold microelectrodes in an acetonitrile solution. The oxidation step is carried out by an electrochemical reaction in which the gold electrode in the center (the gate electrode) of the polymer layer is given a positive potential with respect to a reference electrode in solution. As the polymer impedance drops a current may pass between the two other electrodes, which are placed 20 microns apart. These electrodes are designated as the source and the drain.

The justification for calling the device a "transistor" is that a small signal (which oxidizes the polymer) is used to induce a large signal between the source and the drain.

The basic arrangement used by Wrighton and co-workers is shown in Fig. 23. The silica layer is 0.5 micrometers thick. The source, gate, and drain electrodes are of gold, 3 μm wide, 140 μm long and 0.1 μm thick. The polypyrrole present has a thickness of several hundred Å.

One of the disadvantages of Wrighton's device is the 10 second on-off switching time (63), but this may be overcome by the use of other types of material.

More recently Wrighton published (64) a similar study using polyaniline on a two electrode apparatus (Fig. 24). Instead of a gate electrode to control the polymer conductivity, proton doping is used. Polyaniline must be protonated to achieve high conductivity. Thus, the device acts as a pH (proton) sensor. This method could likely be extended for use as a chemical sensor, where the chemical being monitored provides a signal by affecting the source to drain current.

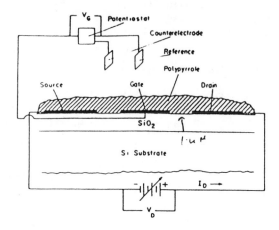

Figure 23.   Transistor composed of polypyrrole.

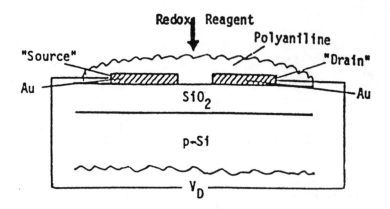

Figure 24.   A chemical sensor composed of polyaniline on two Au microelectrodes.

## 11.   NEEDED RESEARCH IN THE ELECTROCHEMISTRY OF ELECTRONICALLY CON-DUCTING POLYMERS

### 11.1   Conductivity

Our ability to rationally choose and make polymers which will be electronically conducting is poor.   Hence, the most important research topic is to investigate why polymers conduct electronically.

In this area there are many pathways which could attract research. The first one is the most simple: A widespread search for electronically conducting polymers at the empirical level. The field of polymers itself is of great breadth. It would be enlightening to take typical examples of polymer classes and examine their electronic conductivity in the normal, oxidized and reduced state, in and out of solution, and over a wide range of temperatures.

Related to this area is the question of doping as it is related to oxidation/reduction reactions. Thus, there are few polymers (e.g., polysulfurnitride) which conduct electronically without the induction of an oxidation/reduction process. Can organic polymers be developed which do not require charge injection to give high conductivity?

The third aspect concerns the effect of the physical arrangement of the polymer strands when conducting. The present picture is one of tortuosity among the strands and it is possible to see that a straightening out of such strands so that they grow parallel could give rise to an increase of conduction.*

It is interesting to note here the contributions recently made by D. W. Bennett and his student, J. S. Rommel (66) in the new series of organic bridging groups. Diisocyanobenzene $(CN-C_6H_4-NC)$ is used to link two transition metals. These systems can possibly be joined together to form an electronically conducting polymer.

## 11.2 The Surface

Compared to the internal properties of electronically conducting polymers, the surface is virtually unstudied, particularly the solid-liquid interface. Among the measurements which should be made include the degree of roughness and the characterization of amorphousness.

Topographical aspects of the microstructure of the polymers will be of interest. For example, microelectrodes are finding increasing use in electrochemistry. These usually have tips in the 100 Å region. In scanning tunneling microscopy, the phrase "1 Å tip" is used. To what extent the protruding organic groups on the surface give rise to micro-electrodes? Thus, the microstructure of such polymer surfaces at the 10 Å level, would be of importance in relating electrochemical reactivity to the dominant effect of the groups (Fig. 28). Scanning tunneling microscopy along with photoelectron spectroscopy, Auger, ESCA, etc., would be helpful in understanding the surface structure. Adelmann (67) at the University of Hawaii has recently developed x-ray techniques which allow one to look at the surface in solution.

---

*Recent work (65) of Martin, et al., at Texas A&M University, has shown that the growth of polymers through small pores can give rise to a straightening out to a degree and is accompanied by an increase in the polymer reduction rate (Figs. 25-27).

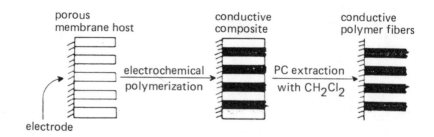

Figure 25.  The electronically conductive composite route to
fibrillar polymer morphology.

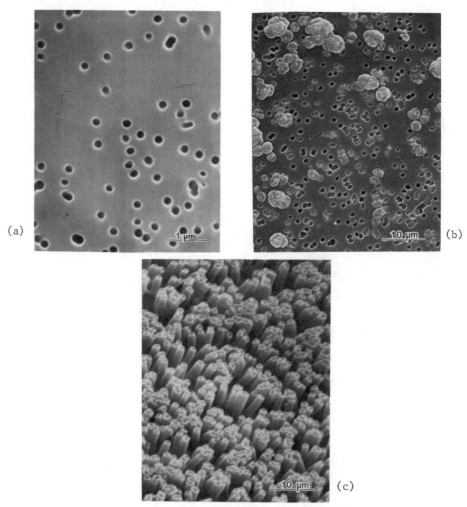

(a)

(b)

(c)

Figure 26.  SEM of Nucleopore film, Nuclepore with polypyrrole,
and polypyrrole fibrils.

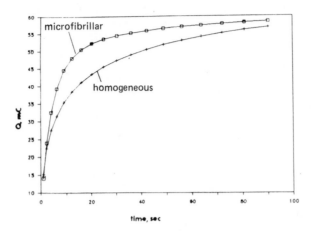

Figure 27. Charge vs. time transients for reduction of
microfibrillar and homogeneous polypyrrole films (Q = 0.60C).

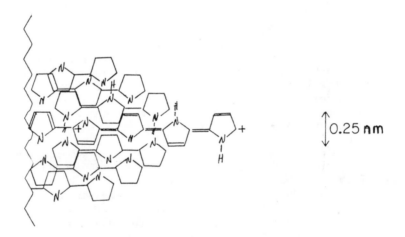

Figure 28. Schematic of a possible microelectrode tip
employing polypyrrole.

11.3 The Double Layer at the Polymer-solution Interface

Studies of the semiconductor-solution interface (68) are quite rare and
the ideas of the structure of so-called insulator-solution interfaces
(69) are just beginning.

Investigation into the polymer-solution interface would involve,
to begin with, the determination of the energy gap; the flatband potential; the carrier concentration; and the carrier mobility (by Hall
effect measurements).

Thereafter, the double layer itself can be looked at in terms of these results along with impedance measurements. These can be interpreted in respect to an estimate of the location of the potential drops in the double layer region, possibly the most important thing in respect to obtaining a model structure of the double layer.

Thus, if the semiconductor corresponds essentially to an insulator of the Schottky barrier type, use of Mott-Schottky plots will allow the determination of the capacitance of the inner layer. Utilization of impedance measurements with different frequencies may give rise to the possibility of determining the double layer capacity separate from the inner layer. In this way a map of the double layer, an estimation of the Helmholtz potential difference and the potential difference (pd) in the space charge region may be obtained. The pd in the Helmholtz layer is, however, not only given by the charge on the surface of the polymer, but also by the potential difference due to aggregated layers which form within it, and in particular the solvent dipole layer (70).

Whether the location of the potential difference at the interface corresponds to Fig. 10 or Fig. 11 will depend upon the location of the charge in the semiconductor. The possibility that the individual fibrils and threads from the polymer have to be treated as individual macromolecules must also be taken into account in interpretation of any double layer data.

A big area of study in the double layer is the measurement of adsorption. Such measurements are concerned with the degree of the surface covered with adsorbed material (organics or ions - the solvent is always adsorbed) as a function of concentration of the material in solution. This is the "adsorption isotherm," but there is also another typical measurement which is made in electrochemical situations of adsorption, and that is the connection between the coverage (theta) and the potential of the surface as a function of potential, i.e., the theta-V plot.

What would these studies give us? They give a map of the fields which pertain to the interface. In the case of semiconducting polymers it is likely that the double layer region starts back in the interior (perhaps 100 to 1000 Å) and continue through the surface with a high field strength near to the interface with the solution followed by (depending on the concentration of the ionic solution) a gentler change of potential with distance into the solution.

## 11.4  Kinetics

It is easier to see what experiments should be done in the electrode kinetic regime. Some basic electrode kinetics at polymer-solution interfaces must be measured - and this has not been successfully done. Thus, with the typical redox reactions of the ferrous/ferric type, it is important to establish Tafel parameters, particularly the parameter beta, and exchange current densities and rate constants.

The second kind of electrochemical reactions which should be carried out at some polymer-solution interfaces are those connected with proton discharge reactions and oxygen reduction. The oxygen reduction reaction at pH = 0 has a standard potential of 1.23 volts and the area of potential in which the reduction of oxygen occurs is 0.6-0.8 volts positive to the hydrogen electrode. This means that the surface is enormously oxidative in spite of the fact that oxygen is being reduced, and therefore whether there are polymer surfaces which will stand up to this degree of oxidation and provide interesting surfaces for oxygen reduction is unknown and should be of great interest, particularly in the area of fuel cell electrochemistry.

Last of all, use of polymers as electrodes should be oriented toward the bioelectrochemical direction. One interesting possibility in the use of polymer electrodes is the study of cytochrome C, which is now being studied at metal electrodes (71).

## 11.5  Corrosion of Polymer Electrode Properties

The promise of electronically conducting polymers as electrodes depends upon their stability, and no systematic examination has hitherto been given in this area.

The first thing needed is a test-bed experiment and some form of the Stern-Geary experiment would be of interest. This could be automated so that a polymer placed in solution and allowed to find a steady state could then be subjected to anodic and cathodic pulses. This would be used to indicate the corrosion rate.

This rapid quantitative scan of corrosion could then be supplemented by the introduction of a simultaneous examination of the surface by two optical means. There is no doubt that ordinary optical microscopy with magnifications at 500 times or so, and the involvement of a microscope with an immersion eyepiece can be of interest in finding out what happens in the more violent corrosion reactions. Interferometry may be used to give an idea as to when a surface is retreating by a rate of say 100 A per hour.

Finally, it may be pointed out that some experiments on stress may be necessary to determine the mechanical properties of polymers as a function of their existence in solution and as a result of the reaction being carried out there. Such experiments are necessary because much corrosion is internal and not apparent on the surface. The development of spreading cracks within the surface is what gives rise to a lack of strength in metals and something similar may be happening in a polymer. In this case it would be possible to make stress-strain curves on polymers and to carry out electrode reactions and observing how the stress-strain curve is potential dependent.

## 12.  SUMMARY

The origin of the field concerned is traced back to Kallmann and Pope (1) of 1960 who showed that the insulator, anthracene, could become conducting in contact with ionic solutions. The discovery by MacDiarmid (5a) that doping of polyacetylene could lead to various degrees of electronic conductivity began the modern phase of the work. Electrochemical applications appear to have a good prospect. The MacDiarmid work was in 1977, and by 1981 reports of batteries (51) were published.

Diaz (55) has been prominent in broadening the field out beyond polyacetylene.

The scope of the electrochemical applications is outlined.  The
availability of a large range of electronically conducting organic compounds would multiply the scope of available materials in electrochemistry.  Applications to electrocatalysts would give the possibility of
lowering the polarization, hence increasing the efficiency, of electrochemical energy converters, thereby lowering cost.  The lightweight of
prospective polymer batteries brings with it the possibility of utilizing electrochemical power sources more widely, perhaps in air vehicles.
The extension of the use of the materials concerned to biochemistry and
biomedicine, particularly prosthetics, seems possible.  Biosensors
should be stressed, e.g., automated prosthetic insulin producers or
artificial pancreatic organs could be developed.

A swathe of other applications is briefly surveyed:  Fuel cells;
solar panels; devices with no IR detectability (e.g., for the electrical
supply of tanks); cheap photovoltaics; prosthetics; and SDI applications
where lightweight electronic devices for lift-off into space are needed.

The basic mechanism of the conductance is easily understood in a
qualitative sense by the fact that reduction or oxidation reactions
cause the polymer to have an excess of electrons or holes over that
which it would have in the absence of such reactions.  However, there
is little ability to predict in any detail the relationship between the
structure of a compound and its electronic conductivity.

The double layer structure at the electronic conducting polymer
is discussed theoretically (in the absence of experimental work).  There
are two limiting cases.  In one the double layer region is entirely at
the solid-liquid interface.  In the other it extends into the organic
solid by varying degrees, perhaps as far as 1000 $\overset{\circ}{A}$.  In such a case,
the pd across the solution boundary varies little and the potential
determining the electrochemical reaction rate occurs within the
semiconductor (Schottky-approximation).  The existence of surface
states may make electronically conducting polymers into bodies where
the double layer resembles more the metal-solution interface than
that associated with the Schottky semiconductor.

Photoelectrochemical converters have  been little stressed, but
recent Japanese work (50) seems to offer an efficiency of about half
the best obtained with inorganic materials but at a cost of perhaps 1%.

In the development of batteries to date, the most notable are
those containing n and p type polymer, e.g., the perchlorate-doped polyacetylene in conjunction with the lithium doped polyacetylene, and
acetonitrile.  Aqueous polyacetylene battereis are under construction.
Theoretical work has been done here by Will (56), who has derived
equations for the slow change of a battery potential resulting from
dopant diffusion within the solid.  Mermilliod and co-workers (59) have
demonstrated that much of the electrical capacity of the batteries arise
not because of the conversion of chemical to electrical work as with
normal batteries, but because of the storage of electricity in the
double layer, very large because of the high surface area to bulk ratio
in many polymers.

Molecular electronic devices offer strong possibilities for application.  Wrighton's work (62-64) on a transistor has recently been published.

Research needs in the electrochemical areas are surveyed:

1.  Most important is the need to increase the predictability of the theories applied to conductivity and to apply this to the invention of increased numbers of conducting polymers.  Simple empirical work applied to the large number of existing polymers (with doping) is suggested.

2.  The surface chemistry of electronically conducting polymers represents a virgin field.  The surface electrochemical area, particularly study of the double layer and its interface (metallic or Schottky?) awaits development.  Electrodekinetics has not been examined from the basic point of view.  For example, it is not known to what degree diffusion and transport in the pores of polymer electrodes control the rate of electrode reactions.  What is the order of magnitude of rate constants involved in various electrochemical reactions which have electronically conducting polymers as their substrate?  To what extent is the energy storage capacitative?  Corrosion and material properties are unexplored.

## 13.  PRINCIPAL CONCLUSIONS

1.  There is a need for the development of a wider series of electronically conducting polymers.

2.  A big hindrance in the field is the absence of a predictive theory of conductivity.

3.  There is a remarkable absence of any studies of the electrical character of the 100 Å around the interface between electrode and solution.

4.  Of the prospective applications, biosensors and the biomedicine direction appear to have great potential.

5.  Use of conducting polymers as photoelectrodes has received little attention.  Efficiencies obtained with polyacetylene are around half those obtained with silicon.

6.  Batteries show promise.  The stability of the batteries is still too low for widespread commercial use.

7.  Transistors have already been made using electronically conducting polymers.  However, where interfaces involving solutions are concerned the speed of the switching has to be improved.

REFERENCES

1.  H. Kallmann and M. Pope, J. Chem. Phys. 32,300 (1960).

2.  H. Mehl, J. M. Hale, and F. Lohmann, J. Electrochem. Soc. 113 (11), 1166 (1966).

3.  A. A. Bright et al., Phys. Rev. Lett. 34(4), 206 (1975).

4.  R. J. Nowak et al., J. C. S. Chem. Comm. 1977, 9.

5.  C. K. Chiang et al., Phys. Rev. Lett. 39(17), 1098 (1977); A. J. Heeger and A. G. MacDiarmid, The Physics and Chemistry of Low Dimensional Solids, L. Alcacer, ed., D. Reidel Pub. Co., Boston, 1980, 353-391; ibid. 393-402; P. J. Nigrey, et al., J. Electrochem. Soc. 128, 1651 (1981); P. J. Nigrey, et al., Mol. Cryst. Liq. Cryst. 83, 309 (1982); W. Wanqun, R. J. Mammone, and A. G. MacDiarmid, Syn. Metals 10, 235 (1985).

6.  R. Pethig and Szent-Gyorgyi, Proc. Natl. Acad. Sci. 74, 226 (1978).

7.  J. O'M. Bockris, F. Gutmann, and M. A. Habib, J. Biol. Phys. 13, 3 (1985).

8.  J. O'M. Bockris and A. K. N. Reddy, Modern Electrochemistry Plenum Press, New York, 1970.

9.  W. Mehl and J. M. Hale, Farad. Soc. Disc. 45, 30 (1968).

10. A. A. Khatiashvili and L. I. Boguslavski, Elektrokhimiya, 11, 1635 (1975).

11. A. F. Diaz and J. I. Castillo, J. C. S. Chem. Comm. 1980, 397.

12. R. A. Bull, F. R. Fan, and A. J. Bard, J. Electrochem. Soc. 129 (5), 1009 (1982).

13. A. F. Diaz and J. A. Logan, J. Electroanal. Chem. 111, 111 (1980).

14. R. J. Waltman, J. Bargon, and A. F. Diaz, J. Phys. Chem. 87, 1459 (1983).

15. S. Srinivasan, et al., 'Electrochemistry in the Biomedical Sciences' in Electrochemistry, ed. by Bloom and Gutmann, Plenum Press, New York, 1977, p. 57.

16. J. L. Bredas, R. R. Chance and R. Silberg, Mol. Cryst. Liq. Cryst. 77, 319 (1981).

17. J. L. Bredas, B. Themans, J. M. Andre, R. R. Chance, D. S. Boudreaux and R. Silbey, J. Phys. Coll. 44 (6), C3-373 (1983).

18. P. M. Grant and I. P. Batra, Syn. Met. 1, 193 (1979/80).

19. W. P. See, J. R. Schrieffer, and A. J. Heeper, Phys. Rev. Lett. 42 (25, 1698 (1979).

20. P. L. Kronick, H. Kaye, E. F. Chapman, S. B. Mainthia, and M. M. Labes, J. Chem. Phys. 36, 2235 (1962).

21. A. J. Heeger and A. G. MacDiarmid in The Physics and Chemistry of Low Dimensional Solids, L. Alcacer ed., D. Reidel Pub. Co., Boston, 1980, pp. 353-391.

22. A. G. MacDiarmid and T. C. Chung, J. Phys. Coll. 44 (6), C3-513 (1983).

23. J. C. Scott, P. Pfluger, M. T. Kroumbi, and G. B. Street, Phys. Rev. B. 28 (4), 2140 (1983).

24. J. C. Scott, J. L. Bredas, K. Yakushi, P. Pfluger, and G. B. Street, Syn, Met. 9, 165 (1984).

25. J. L. Bredas, B. Themans, J. M. Andre, R. A. Chance, and R. Silbey, Syn. Met. 9, 265 (1984).

26. F. Wudl, Pure and Appl. Chem. 54 (5), 1051 (1982).

27. J. L. Bredas, B. Themans, J. M. Andre, A. J. Heeger, and F. Wudl, Synthetic Metals 11, 343 (1985).

28. R. A. Ball, F-R. Fan and A. J. Bard, J. Electrochem. Soc. 129 (5), 1009 (1982).

29. J. O'M. Bockris and M. S. Tunulli, Bioelectrochemistry, ed. by H. Kaiser and F. Gutmann, Plenum Press, N. Y. 1980, p. 19-29.

30. M. R. Tarasevich in Comprehensive Treatise of Electrochemistry, v. 10, ed. by Srinivasan, Chizmadzlev, Bockris, Conway and Yeager, Plenum Press, N. Y., 1985.

31. P. N. Sawyer and S. Srinivasan, J. Colloid. Interface Sci. 32, 456 (1970) and in Comprehensive Treatise of Electrochem. v. 10, 1985.

32. S. Srinivasan, G. Cahan, Jr., G. Stoner, Electrochemistry: The Past Thirty and the Next Thirty Years, ed. by H. Bloom and F. Gutmann, Plenum Press, N. Y. 1977, pp. 57-84.

33. J. Albery, B. Haggett, and D. Snook, New Scientist, Feb. 13, 1986, 38.

34. W. J. Alberg, P. N. Bartlett, and D. H. Cranston, J. Electroanal. Chem. 194, 223 (1985); E. F. Bowden, F. M. Hawkridge, and H. N. Blount, J. Electroanal. Chem. 161, 355 (1984).

35. I. V. Berezin and A. A. Klesov, Russian Chem. Rev. 45 (2), 87 (1976); M. R. Tarasevich, J. Electroanal. Chem. 104, 587 (1979); G. G. Guilbault in Medical and Biological Applications of Electrochem. Devices, ed. by J. Koryta, John Wiley & Sons, LTD., 1980; J. M. Laval, C. Bourdillon, and J. Moiroux, J. Am. Chem. Soc. 106, 4701 (1984).

36. E. Gileadi, E. Kirowa - Eisner, and J. Penciner, Interfacial Electrochemistry:  An Experimental Approach, Addison-Wesley Pub., Reading, Mass., 1975.

37. J. O'M. Bockris and S. U. M. Khan, J. Electrochem. Soc. 132 (11), 2648 (1985).

38. K. Chandrasekaran and J. O'M. Bockris, submitted for publication.

39. M. A. Habib and J. O'M. Bockris, J. Electrochem. Soc. 132 (1), 108 (1985).

40. T. A. Skotheim, S. W. Feldberg, and M. B. Armand, J. Phys. Coll. 44 (6), C3-615 (1983).

41. S. W. Feldberg, J. Am. Chem. Soc. 106, 4671 (1984).

42. K. H. Dietz and F. Beck, J. Appl. Electrochem. 15, 159 (1985).

43. P. Audebert and G. Bidan, J. Electroanal. Chem. 190, 129 (1985).

44. R. C. M. Jakobs, L. J. Janssen, and E. Barendrecht, Electrochimica Acta 30 (10), 1313 (1985).

45. R. C. M. Jakobs, L. J. J. Janssen, and E. Barendrecht, Electrochimica Acta, 30 (11), 1433 (1985).

46. M. Saloma, M. Aquilar, and M. Salmon, J. Electrochem. Soc. 132 (10), 2379 (1985).

47. R. A. Saraceno, J. G. Pack, and A. G. Ewing, J. Electroanal. Chem. 197, 265 (1986).

48. A. Fujishima and K. Honda, Nature 238, 37 (1972).

49. S. Chen, A. Heeger, Z. Kiss, A. MacDiarmid, S. Gau, and D. Peebles, Appl. Phys. Lett. 36 (1), 96 (1980).

50. H. Shirakawa, S. Okada, M. Aizawa, J. Yoshitake, and S. Suzuki, Synthetic Metals, 4, 43 (1981).

51.  P. J. Nigrey, D. MacInnes, Jr., D. P. Nairns, A. G. MacDiarmid and A. J. Heeger, J. Electrochem. Soc. 128, 1651 (1981).

52.  R. Somoano, Appl. Phys. Comm. 1 (2), 179 (1981-82).

53.  P. J. Nigrey, A. G. MacDiarmid, and A. J. Heeger, Mol. Cryst. Lig. Cryst. 83, 309 (1982).

54.  R. B. Kaner, A. G. MacDiarmid and R. J. Mammone, Polymers in Electronics, ed. by T. Davidson, ACS Pub., Washington, 1984.

55.  R. J. Waltman, A. F. Diaz and J. Bargon, J. Electrochem. Soc. 131 (6), 1452 (1984).

56.  F. G. Will, J. Electrochem. Soc. 132 (10), 2351 (1985).

57.  A. G. MacDiarmid, S-L. Mu, N. Somasiri, and W. Wu, Mol. Cryst. Liq. Cryst. 121, 187 (1985).

58.  A. Kitani, M. Kaya, and K. Sasaki, J. Electrochem. Soc. 133 (6), 1069 (1986).

59.  N. Mermilliod, J. Tanquy and F. Petiot, J. Electrochem. Soc. 133 (6), 1073 (1986).

60a. Molecular Electronic Devices, ed. by F. L. Carter, Marcel Dekker, Inc. N.Y. 1982;
  b. Yates, F. E. (1984)  Report on Conference on Chemically-based Computer Designs.  Crump Institute for Medical Engineering Report CIME TR/84/1, University of California, Los Angeles, CA  90024.

61.  B. Lovrecek, A. Despic, and J. O'M. Bockris, J. Phys. Chem. 63, 750 (1959).

62.  H. S. White, G. P. Kittlesen, and M. S. Wrighton, J. Am. Chem. Soc. 106, 5375 (1984).

63.  G. P. Kittlesen, H. S. White, and M. S. Wrighton, J. Am. Chem. Soc. 106, 7389 (1984).

64.  E. W. Paul, A. J. Ricco, and M. S. Wrighton, J. Phys. Chem. 89, 1441 (1985).

65.  R. Penner and C. Martin, Submitted for publication.

66.  Chemical and Engineering News, Sept. 30, 1985, p. 22.

67.  Ehrich, Electron Spectroscopy Theory, Technique and Applications, v. 3, pp. 1-39.

68. Y. V. Pleskov, Comprehensive Treatise of Electrochem., v. 1, Plenum Press, N.Y. 1980, pp. 291-328.

69. L. I. Boguslavski, ibid., pp. 329-331.

70. M. A. Habib and J. O'M. Bockris, Langmuir, in press.

71. E. F. Bowden, F. M. Hawkridge, and H. N. Blount, J. Electroanal. Chem. 161, 355 (1984).

# IN SITU RAMAN EXPERIMENTS ON POLYACETYLENE IN ELECTROCHEMICAL CELLS

S. LEFRANT
Laboratoire de Physique Cristalline,* Université de Nantes
2, rue de la Houssinière
44072 NANTES CEDEX 03
France

ABSTRACT. The use of conducting organic polymers as electrode materials in rechargeable batteries has been demonstrated in the past few years and in particular, polyacetylene, the prototype polymer, can be reversibly oxidized or reduced by electrochemical procedures. These doping processes lead to the incorporation of counterions within the polymer with a concomitant increase in the electrical conductivity into the metallic regime. This technique is now widely applied for a number of organic polymers used as electrodes since it offers a more precise control of the doping level as well as a much better dopant homogeneity than for a chemical doping process.

In situ Raman experiments are presented for an n-type doped polyacetylene film as a function of the doping level and compared to p-type doped systems. The modification occuring on the film during a charge-discharge cycle can be analyzed in details and in particular, the appearance of new features on the Raman spectra can be interpreted in terms of vibrational modes induced by doping as already evidenced in infrared spectroscopy. The possibility of using the Raman spectroscopy to study the morphology of the sample is described and the information gained from in situ experiments during the electrochemical process on electrodes composed of conducting organic polymers is emphasized.

## 1. INTRODUCTION

The usefulness of Raman spectroscopy in the study of conducting polymers has been clearly demonstrated these last years, especially in the case of polyacetylene, $(CH)_x$, which has been the subject of numerous papers (1). First, the two isomers, cis-and trans-$(CH)_x$ induce completely different Raman spectra and therefore, any isomerization process can be followed in details by using this spectroscopic technique. Then, trans-$(CH)_x$ exhibits a peculiar behavior when the laser excitation line is changed from the red range (typically 676.4 nm) to the violet (457.9 nm) and even to the U.V. region (351.1 nm). In addition to the two main Raman bands peaked at 1060 and 1458 $cm^{-1}$, satellite components develop by using an excitation energy of high value. Different interpretations

37

L. Alcácer (ed.), Conducting Polymers, 37–46.

have been given for this phenomenon : distributions of conjugate segments (2-4), hot luminescence (5), and more recently, amplitude modes of
the Peierls distortion (6). The purpose of this paper is not to discuss
the validity of one particular model with respect to the others but to
emphasize the capability of the Raman technique to provide important information  on the state and on the morphology of the polymer.

This technique may be used as well in the case of doped polyacetylene in which electrons are transferred from or to the carbon chains by
dopants. One of the main problem arising when samples are doped chemically comes from the difficulty to control carefully the doping process -
especially in the low concentration range - and to be sure of the dopant
homogeneity. Moreover, because of the very high absorbance of the polymer films, the Resonance Raman Scattering (RRS) is able to investigate
mainly the surface of the sample and one may deal with a dopant concentration very different from what can be determined in average by weight
uptake  for example. A considerable improvment was made when it was demonstrated that polyacetylene can be controllably oxidized or reduced
electrochemically (7) with the incorporation of counterions. Polyacetylene films can be used as the cathode or the anode materials in rechargeable batteries. This offers the possibility to perform "in situ" Raman
experiments, provided a glass or quartz window is inserted to the cell
in an appropriate  position. The advantages consist in an accurate determination of the dopant concentration by measuring the charges passing
through the electrode, a better control of the equilibrium (constant
$V_{oc}$ voltage), a possibility to perform doping-dedoping cycles. All these
parameters turn out to be important in the study of the morphology of
$(CH)_x$ films.

In this paper, we mainly describe "in situ" experiments on n-type
doped samples. A comparison is made with previous similar experiments
on p-type doped polymers (8). Recent experiments performed on other conducting polymers such as polyaniline (9) are also briefly described in
the last part of this paper.

## 2.    EXPERIMENTAL DETAILS

Most of the studies reported here on n-typed systems were performed
on electrochemical cells built at the University of Pennsylvania according to a technique already efficiently tested (10).  One electrode was
composed of a cis-rich $(CH)_x$ film pressed onto a Pt grid current collector with an attached wire lead. The second electrode was made of a Li
plate. The Li was embedded in a nickel mesh with an attached nichrome
wire lead. Electrodes were separated with a piece of kiln-dried glass
filter paper. The entire assembly was inserted into a glass cell
(3 x 9 x 40 mm), which was sealed after approximately 0,5 ml electrolyte was added (1.0 M Li Cl $O_4$/THF in the case of n-type dopants). Another
type of electrochemical cell, with a second Li electrode as a reference
and an additional vessel used to pump cryogenically the electrolyte was
sometimes used.

Raman spectra were recorded at room temperature using standard equipment and a 90° scattering geometry configuration was adopted. Laser lines were provided by either an argon or a krypton cw laser. A dye laser

was also used from time to time whenever the cis($CH)_x$ content determination was wanted. All spectra were taken "in situ" in the cell and additional background scattering was minimized by setting the cell in an appropriate position.

## 3. RESULTS

Let first recall briefly the main features of the RRS spectra in polyacetylene. Cis-$(CH)_x$ is characterized by three main lines at 908, 1247 and 1541 $cm^{-1}$ and a maximum intensity is observed when $\lambda_L \simeq 600$ nm. Going out of this resonance weakens Raman spectra, but no (or very little) shift in the frequencies is detected.

Trans-$(CH)_x$ exhibits two main lines at 1064 and 1458 $cm^{-1}$ when the exciting wavelength is tuned at the maximum of the absorption. Two weaker ones are seen at 1015 and 1290 $cm^{-1}$ in this spectral range and a third one at 2990 $cm^{-1}$. Except the one at 1015 $cm^{-1}$ (assigned to an infrared mode becoming active by symmetry breaking in short segments), the four others are predicted by a normal modes analysis (11). When the laser line is now set to a lower wavelength, i.e. 457.9 nm, satellite components grow up at $\simeq$ 1130 and 1520 $cm^{-1}$ respectively, whose intensity with respect to the main bands depend on the sample. All these results have been described in details by several groups (3, 12). As an interpretation of this behavior, we choose the bimodal distribution model developped in (3) and showed that it was appropriate to describe RRS spectra in many different cases : samples with different cis/trans contents, films coming from different synthesis. In the case of doped systems, providing we assume that Raman scattering is only sensitive to the undoped parts of the polymer (13), the results indicate a general shortening of the trans-conjugated segments. This is evidenced by i) a upwards shift in frequencies of the two main bands with bandshapes becoming more symmetrical when $\lambda_L = 676.4$ nm, ii) the disappearance of the low frequency components when $\lambda_L = 457.9$ nm indicating that the concentration of long conjugated segments decreases subtantially. When the concentration level increases, an additional band is observed at $\simeq$ 1600 $cm^{-1}$. These features, described in (13) and further analyzed in (4), mainly characterize p-doped systems. In the case of n-doped polyacetylene, fewer results are available in the literature. (CH $Li_y)_x$ prepared chemically is the most studied sample (14). Although the electrolyte solution was well defined ($10^{-2}$ M/$\ell$ solution of Li-benzophenone in THF), the doping time was taken as a measurement of level of the doping process. In figure 1, taken from ref. 14, Raman spectra of cis-$(CH)_x$ doped with Li are presented for $\lambda_L = 600$ nm. It is observed that the intensity of the "cis" lines decreases and Raman bands characteristic of the trans isomer appear. After a long time, the "cis" lines are no longer detected and the spectrum is composed of "trans" bands at $\simeq$ 1080 and 1465 $cm^{-1}$. In addition, small additional bands begin to appear at $\simeq$ 1271 and 1580 $cm^{-1}$ (dashed in fig. 1d). In Fig. 2, we present Raman spectra induced for a highly doped sample in which the lines assigned to the trans-$(CH)_x$ isomer are very weak (1140 and 1548 $cm^{-1}$ in spectrum 2a). In addition, two strong bands are now recorded at 1280 and 1600 $cm^{-1}$ (for $\lambda_L = 457.9$ nm).

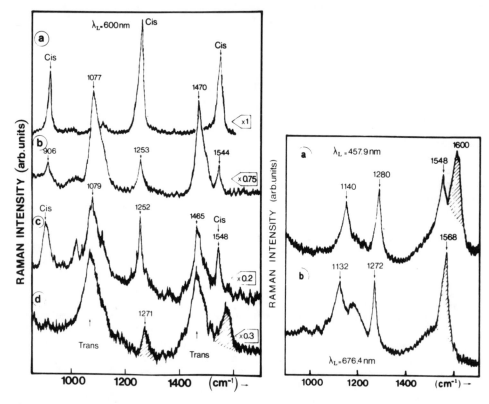

Figure 1. Raman spectra of cis-(CH)$_x$
doped with Li ; $\lambda_L$ = 600 nm ;
T = 295 K
a) before doping ;
b) after 2 mn doping ;
c) after 7 mn doping ;
d) after 13 h doping .

Figure 2. Raman spectra of cis-(CH)$_x$
highly doped with Li ; T = 295 K
a) $\lambda_L$ = 457.9 nm ;
b) $\lambda_L$ = 676.4 nm .

It appears from the preliminary study on chemically doped samples that a
strong modification occurs upon doping with electron donors. Besides the
cis-trans isomerization, for which the dopant level is not known, a dras-
tic change in the morphology or in the structure of the polymer film is
observed and needs to be studying in details.

     For this purpose, the electrochemical procedure turns out to be real-
ly appropriate, provided "in situ" experiments are performed since n-doped
(CH)$_x$ films are very sensitive to the ambient atmosphere. In the neutral
state, the electrochemical cell was completely stable (The $V_{OC}$ voltage :
Li/Li Cl O$_4$ (THF)/(CH)$_x$ was typically 2.3 V). The n-doping of the (CH)$_x$
film was achieved by lowering the potential of the cell (potenstiostat/
galvanostat PAR model 273).

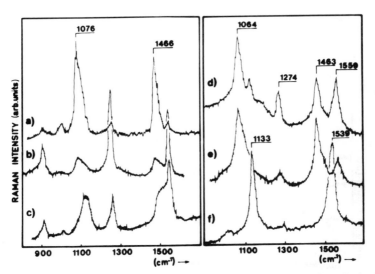

Figure 3. Raman spectra at RT of electrochemically Li-doped $(CH)_x$

— $(CH\ Li_{0.004})_x$ : a) $\lambda_L$ = 676 nm ; b)  $\lambda_L$ = 600 nm ; c) $\lambda_L$ = 457.9nm

— $(CH\ Li_{0.036})_x$ : d) $\lambda_L$ = 676 nm ; e)  $\lambda_L$ = 600 nm ; f) $\lambda_L$ = 457.9nm

In Figure 3, Raman spectra are shown for two different concentrations of $Li^+$ and for different excitation wavelengths. Starting with a cis-$(CH)_x$ sample, it is found that $(CH\ Li_{0.004})_x$ has not been isomerized yet to the trans isomer (refer for example to curve b in figure 1 for $\lambda_L$ = 600 nm). At the same time, the trans bands have peaks located at 1076 and 1466 cm$^{-1}$, characteristics of rather short conjugated trans segments. For a higher dopant concentration, i.e. 3,6 %, no "cis" lines are observed indicating a full isomerization. The "trans" bands have maxima measured at 1064 and 1463 cm$^{-1}$ for $\lambda_L$ = 676.4 nm, shifted to 1133 and 1539 cm$^{-1}$ respectively for $\lambda_L$ = 457.9 nm. For this dopant level, it should be noticed that the two additional bands, reported in the case of chemically n-doped systems, are observed at 1274 and 1559 cm$^{-1}$ only for the red excitation laser line ( $\lambda_L$ = 676.4 nm). This is an important result showing, for example, that the change in the structure is first associated with long chains. In the doping is continued to a higher level, the additional bands appear for all excitation wavelengths as seen in the case of chemically doped $(CH)_x$ films.

Another strong advantage of the electrochemical procedure is to give the possibility of performing many doping-dedoping cycles. Raman spectra taken at different stages of such a cycle are shown Figure 4 for $\lambda_L$ = 457.9 nm. Another type of n-dopant was used in this study : $Bu_4N^+$. The starting material contained a mixture of cis and trans isomers and the $V_{oc}$ voltage of the as-built electrochemical cell was 1.87 eV. Lowering the voltage down to 0.95 V led to a Raman spectrum characteristic of a reasonably doped sample, i.e. showing "trans" bands superimposed to

the two additional features described above. Going back to the voltage
corresponding to the neutral state, i.e. 1.55 V then 2.2 V ends up with
a Raman spectrum of an undoped trans-(CH)$_x$ sample.

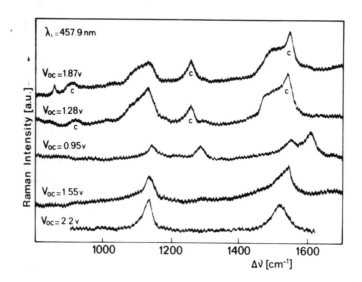

Figure 4. In situ Raman spectra of (CH)$_x$ doped with Bu$_4$N$^+$ at different
          stages of a charge-discharge cycle ; T = 295 K ; $\lambda_L$ = 457.9 nm

The process is therefore completely reversible and a carefull look
on the bandshapes of the Raman bands indicate that the trans isomer can
be considered as a "good quality" sample with respect to the conjugation
length. An evidence for this is given by the double peak structure detected with the blue excitation wavelength when a great care is taken in the
solvent preparation.

Polyacetylene can be electrochemically doped with any alkali-metal as
done chemically. A high doping level gives rise to approximately the
same new features which may depend somewhat on the cation inserted as a
counterion. If we refer for example to the results shown in Fig. 5
(taken from ref. 15), we may notice that the additional bands shift
downwards in frequencies when the cation size increases.

It is not clear at present if this effect is truly due to the cation
size or to different doping levels which were not accurately determined
because the chemical procedure was used. It should be noticed that small
features are sometimes observed at $\simeq$ 960 (see Fig. 5) and also around
1400 cm$^{-1}$.

An important point which may be discussed here concerns the structural rearrangements which may occur after thermal annealings of alkali-
metal/(CH)$_x$ complexes.

Such experiments are associated with an increase in conductivity (16)
and a strong increase in the peak-to-peak width ($\Delta H_{pp}$) of the EPR line (17).

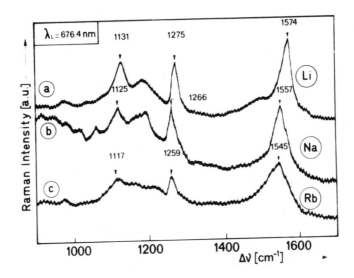

Figure 5. Raman spectra at room temperature induced by alkali-metal
doped polyacetylene for $\lambda_L$ = 676.4 nm
a) $(CH\,Li_y)_x$    ;    b) $(CH\,Na_y)_x$    ;    c) $(CH\,Rb_y)_x$

Raman experiments performed on chemically doped samples revealed only
small modifications upon thermal annealing which consist on a broadening
of the low frequency side of the 1596 $cm^{-1}$ band, likely associated with
a small decrease of the concentration of strongly perturbed doped parts
of $(CH)_x$. This is far from detecting an ordering in polyacetylene as reported by Shacklette et Toth (18) but very recent EPR experiments performed "in situ" on K-doped $(CH)_x$ indicate, after a few cycles,
also some evidence óf structural rearrangements (19). "In situ" Raman
experiments are under way to check this point via the dynamical properties.

DISCUSSION and CONCLUSION

In this paper, we have described Raman experiments performed
"in situ" on n-doped $(CH)_x$ systems prepared electrochemically and compared to results obtained on similar samples chemically prepared (20-23).
By using the electrochemical procedure, it is possible to avoid most of
the difficulties encountered previously such as dopant homogeneity and
concentration level. The possibility of performing charge-discharge cycles
(doping-dedoping) allows a comparison between n-doped and p-doped systems.
In particular, it appears that p-doped systems suffer some degradation
after different cycles (24). Also, Meisterle et al (8), describing
in situ experiments, reported that a high degree of disorder was induced
by doping. These authors mention that a doping-induced isomerized film is
worse than a thermally isomerized one . This is not the case of n-doped
systems which, after dedoping, exhibit the features of a "good quality"

trans-$(CH)_x$ sample (22). It does not look like cycling induces additional disorder. This can be due to the fact that p -dopants, upon cycling, may incorporate the $(CH)_x$ chains trough additional or subtitutional processes.

Raman spectra of lightly n-doped films are interpreted by a shortening of the trans-conjugated segments since the technique is sensitive to the undoped parts of the polymer because of resonance effects. The new feature appearing on highly doped films may have different origins. The band at $\simeq 1600$ cm$^{-1}$, existing also in p-doped films, may be associated to strongly perturbed C = C stretching modes since its frequency is dependent on the exciting wavelength and is therefore associated to the carbon chain. The assignment of the 1270 cm$^{-1}$ is somewhat more speculative. Zannoni et al. (25) calculated a vibrational mode in this range for an undimerized chain. Another possibility is given by the doping-induced infrared modes (26) which could become active by symmetry breaking. It is not clear yet if the $\simeq 900$ cm$^{-1}$ and $\simeq 1400$ cm$^{-1}$ are really detected in highly doped-systems, but recent calculations (27) involving electron-phonon interactions could explain the relative high intensity of the 1270 cm$^{-1}$ mode with respect to the others.

The Raman technique is therefore a very useful tool for studying poly-acetylene used as an electrode in an electrochemical cell. Important information is derived concerning the morphology of the sample. This technique is also applied to electropolymerized conducting polymers such as polyaniline (9) for which different oxidation states exist. In particular, an enhancement of the benzenoïd rings modes in the reduced state and of the quinoïd rings modes in the oxidized states was observed and correlated with absorption measurements. Another type of polymer which may be studied is polyparaphenylene. The vibrational features are now well understood (28). So far, the limitation came from the powder form in which this polymer was obtained by synthesis. Preliminary experiments have nevertheless succeeded (29). Starting materials gave rise, as usual, to a very strong fluorescence. After an electrochemical doping with $PF_6^-$ or $AsF_6^-$, the decrease of the fluorescence was observed together with the appearance of Raman modes characteristic of PPP samples. Experiments must be performed further and the electroreduction synthesis recently applied (30) opens a new development in this way since films on conducting plates are obtained and can be handled more easily.

In conclusion, "in situ" Raman experiments, even if they are far from being easy to run, can give important information relevant to the conducting polymers domain. In the future, they will probably be used as a complementary technique to the other spectroscopic ones already widely used in solid state research implying electrochemical dopings or synthesis.

ACKNOWLEDGEMENTS


I am very much indebted to Drs. MacDiarmid, R.B. Kaner, P. Bernier, F. Rachdi, E. Faulques, A. Chentli and S. Krichene who have participated in some work reported in this paper.


✱ The "Laboratoire de Physique Cristalline" is Unité Associée au CNRS n° 802

1. See for example, papers devoted to this subject in the Proceedings of the International Conference on Physics and Chemistry of Low Dimensional Synthetic Metals, Abano Terme (Italy) (1984), Mol. Cryst. Liq. Cryst. 117, 1-486 (1985)

2. L.S. Lichtmann, Ph.D Thesis, Cornell University, 1981, unpublished

3. H. Kuzmany, Phys. St. Sol. (b) 97, 521 (1980) and H. Kuzmany, E.A. Imhoff, D.B. Fitchen and A. Sarhangi, Phys. Rev. B 26, 7109 (1982)

4. G.P. Brivio and E. Mulazzi, Phys. Rev. B 30, 676 (1984) and S. Lefrant, E. Faulques, G.P. Brivio and E. Mulazzi, Sol. St. Comm. 53, 583 (1985)

5. E.J. Mele, Solid St. Comm. 44, 827 (1982)

6. Z. Vardeni, E. Ehrenfreund, O. Brafman and B. Horovitz, Phys. Rev. Lett. 51, 2326 (1983)

7. See for example, P.J. Nigrey, D. MacInnes Jr., D.P. Navins, A.G. Macdiarmid, A.J. Heeger in Conducting Polymers, Ed. by R.B. Seymour, Plenum Publishing Corp., p. 227 (1981)

8. P. Meisterle, H. Kuzmany and G. Nauer, Phys. Rev. B 29, 6008 (1984)

9. H. Kuzmany and N.S. Sarifici, in press in Synth. Metals (1986), Proceedings of the International Conference on Synthetic Metals, Kyoto (Japon) 1986

10. K. Kaneto, M. Maxfield, D.P. Navins, A.G. Macdiarmid and A.J. Heeger, J. Chem. Soc. Faraday Trans. I 78, 3417 (1982)

11. H. Shirakawa, T. Ito and S. Ikeda, Polym. J. 4, 460 (1973)

12. S. Lefrant, L.S. Lichtmann, H. Temkin, D.B. Fitchen, D.C. Miller, G.E. Whitwell II and J.M. Burlitch, Sol. St. Comm. 29, 191 (1979) S. Lefrant, J. Physique (Paris) 44, C3-247 (1983)

13. E. Faulques and S. Lefrant, J. Physique (Paris) 44, C3-337 (1983)

14. E. Faulques, S. Lefrant, F. Rachdi and P. Bernier, Synth. Metals 9, 53 (1984)

15. S. Lefrant, E. Faulques and A. Chentli, Springer Series in Solid State Sciences 63, 122 (1985)

16. R.L. Elsenbaumer, P. Delannoy, G.G. Miller, C.E. Forbes, N.S. Murphy, H. Eckhardt and R.H. Baughman, Synth. Met. 11, 251 (1985)

17. F. Rachdi and P. Bernier, Phys. Rev. B 33 (1986)

18. L.W. Shacklette and J.E. Toth, Phys. Rev. B 32, 5892 (1985)

19. P. Bernier and C. Fite, private communication

20. Y. Furukawa, I. Harada, M. Tasumi, H. Shirakawa and S. Ikeda, Chem. Lett. p. 1489 (1981)

21. H. Eckhardt, L.W. Shacklette, J.S. Szobota and R.H. Baughman, Mol. Cryst. Liq. Cryst. 117, 401 (1985)

22. S. Lefrant, E. Faulques, F. Rachdi and P. Bernier, Mol. Cryst. Liq. Cryst. 117, 377 (1985)
    S. Lefrant, P. Bernier and R.B. Kaner, Jap. J. Appl. Physics 23, L 883 (1984)

23. See proceedings of the ICSM 86, Kyoto : S. Lefrant, E. Faulques, A. Chentli, F. Rachdi and P. Bernier ;
    J. Tanaka, Y. Saito, M. Shineizu and M. Tanaka
    In press in Synth. Metals (1986)

24. G. Wieners, M. Montkenbush and G. Wegner, Ber. Bunsenges Phys. Chem. 88, 935 (1984)

25. G. Zannoni and G. Zerbi, Solid St. Comm. 48, 871 (1983)

26. C.R. Fincher Jr., M. Ozaki, A.J. Heeger and A.G. MacDiarmid, Phys. Rev. B 119, 4140 (1979)

27. E. Mulazzi (private communication)

28. S. Krichene, J.P. Buisson and S. Lefrant, Proceedings of the ICSM 86, Kyoto ; in Press in Synth. Metals (1986)

29. S. Krichene, F. Maurice, S. Lefrant ; unpublished results

30. J.F. Fauvarque, A. Digua, M.A. Petit and J. Savard, Makromol. Chem. 186, 2415 (1985)

# CONDUCTIVE COMPLEXES OF NOVEL PORPHYRIN AND PHENOTHIAZINE POLYMER SYSTEMS

G. Geib, H. Keyzer and K. G. Reimer
Department of Chemistry and Biochemistry
California State University, Los Angeles
5151 State University Drive
Los Angeles, California 90032    U.S.A.

ABSTRACT. Charge transfer complexes of porphyrins, phthalocyanines and phenothiazines have desirable electrical conductive properties. The parent compounds can also function as polymer precursors which in the past often suffered from poor solubility characteristics, limiting their utility in further chemical reactions. Recently, the synthesis of highly soluble metallotetrabenzporphyrin and phenothiazine homologs has made it possible to study the charge transfer characteristics of those compounds and their complexes in a variety of solvents. We wish to report on alternating and direct current measurements of some metal meso-phenyltetrabenzporphyrins, phenothiazinyls, their iodine and TCNQ complexes in the solid and solution state, and ternary complexes of these compounds with polyvinylpyridine. An iodine:phenothiazine-ionene dimer complex exhibited an unusual stoichiometry, and has the lowest resistivity so far detected in iodine-ionene polymers. Well-defined ternary complexes of metallo-organics, $I_2$ and polyvinylpyridines have markedly lower resistivities and activation energies than the parent compounds or the dual component complexes. The structure/conductivity relationship of several of the complexes and ternary polymeric models are discussed.

## INTRODUCTION

Electron transfer from a donor D to an acceptor A molecule site is facilitated, and the mean free path of the carrier is increased, if an intermediate site exists linking D and A. Such effects are frequent, and of crucial importance, in conducting and photochemical systems. A theoretical treatment has recently been given by Fischer et al[1] and by Karl[2]. Very often, a metal atom is involved as the intermediate;[3] this has been studied[4] for triple-decker multi-electron compounds such as bis(cyclopentadienylcobalt)cyclo-octa-tetracene.

The covalent bonding usually existing between the metal ion and its partner(s) involves considerable electron delocalization[5] in ligand orbitals and the resulting complexes exhibit but little change in their geometrical arrangements upon a change in valency. Ternary adducts arise in either solid or liquid solutions.[6] In the solid state, they

47

L. Alcácer (ed.), Conducting Polymers, 47–63.

often involve guest molecules in a host matrix, where the host forms a charge transfer complex.

The host lattice has cavities which are occupied by guest molecules. Intermolecular forces of the order of 0.2 to 0.5 eV bind guest to host.[7] Thus, e.g., the p-tricyanovinyl-N,N-dimethylaniline auto-complex is deep blue.[8] In chloroform, it forms a ternary complex which is deep red. The resulting electronic properties are quite different from those exhibited by the guest-free charge transfer complex (CTC):[7] the resistivity typically drops from, say, $10^9$ to $10^3$-$10^5$ ohm cm, the activation energy of the conductivity drops from, say 0.3 eV to 0.1-0.2 eV and free radicals become evident from ESR as well as from spectroscopic studies. A typical example[9] of ternary adducts are charge transfer complexes between tetracyanoethylene and aromatic donors, doped with guest donors: the guest complexes yield triplet traps in the host complex crystal and are ESR active. The electron trapped is then delocalized over one donor-acceptor pair.[9]

A serious problem facing research into the properties of well-conducting organic semiconductors is presented by the fact that so many are refractory, amorphous, insoluble in all solvents and virtually chemically inert. This statement applies particularly to polymers. As a consequence their structure, and physico-chemical properties, are almost inaccessible by conventional investigation techniques.

Several avenues towards solution of these problems have been attempted. One may introduce functional groups which could improve solubility in ordinary solvents, or make the compound's structure amenable to further chemical reaction or physico-chemical techniques such as nuclear magnetic resonance. Complexation with paramagnetic ions, for for instance, yields proton relaxation effects resulting in selective[10] broadening of signals from protons close to binding sites. Information from model compounds can be very helpful. X-ray studies of single crystals of model TCNQ-polymer compounds showed that the electrical properties depended mainly on the crystal geometry and not on the length of the chains in the polymer.[11] Inevitably one must make assumptions as to structure-electrical property relationships from the analogous characteristics observed.

A resurgence of interest in ionenes has occurred recently with the discovery[12] that the salts of N-alkylaminephenothiazines form micelles in aqueous and organic solvents; and in particular because the structure of the micelles involve the partial overlap of the phenothiazine rings with each other while the alkylamine sidechains all extend on the same side of the micellar stack.[13] We believe that by attaching the phenothiazine rings as pendant groups to a quaternary ammonium skeleton the resulting polymers may well exhibit interesting conductive properties.

Also, in recent years there has been an increasing interest in the conductivities of metalloporphyrin and phthalocyanine solids, polymers and complexes.[14-23] Several axially oriented macrocyclic polymers have shown[15] relatively high conductivities following iodine oxidation. For example, the nickel and iron phthalocyanines are good electrical insulators in the range of $10^{10}$ to $10^{11}$ ohm cm. However, following iodine oxidation at room temperature, dark resistivities are dramatically increased in the region of $10^3$ to $10^0$ ohm cm. Studies have

shown[17-20] that the conductivities of these compounds are very anisotropic as a result of their nature to stack axially in column formation.
Hanack[22,23] has shown that good conductivities can be obtained by
incorporating bridging ligands in the stacked macrocyclic complexes.
This results in having the conduction pathway transmitted through the
bridging moieties. Porphyrin-ligand charge transfer processes have also
been identified[24] by electronic absorption spectra in solution as in
the case of (pyridine)$_2$Fe$^{II}$tetraphenylporphyrin.

In this study we examined the conductivity relationships and thermal
activation energies of several novel phenothiazines and metallotetra-
benzporphyrins complexed with common electron acceptors. We also
studied ternary complexes of these compounds with polyvinylpyridine.


## MATERIALS AND METHODS

The pentyl dimer of 10-(3-methylaminopropyl)phenothiazine (I), Fig. 1,
was prepared[25] beginning with the synthesis for 10-(3-chloropropyl)phe-
nothiazine (II) by the method of Gozlan et al.[26] The colorless oil
(II) so obtained was found to be 80% pure. Monoethylamine was bubbled
into a mixture of (II) 2.0 g (7.3 mmole, 80%) of the colorless oil, II,
25 ml DMF and 10 ml methanol, until the mixture contained about 0.5 g of
the amine.[28] The mixture was heated to 40°C with stirring. While the
temperature was maintained at 40°C for two hours, monomethylamine was
bubbled slowly into the mixture to maintain its concentration. The
solvent was removed by rotary evaporation under vacuum at 80°C. The
resulting amber solid was dissolved in 50 ml of water and extracted
twice with 20 ml diethyl ether. The water layer was adjusted to pH 10
with 50% sodium hydroxide and extracted 3 times with 20 ml diethyl
ether. The ether was removed by rotary evaporation to yield 1.5 g
10-(3-methylaminopropyl)phenothiazine (III). Into a mixture of 25 ml
DMF, 10 ml methanol and 1.3 g (7.4 mmoles) 1,5 dibromopentane was added
1.0 g (3.7 mmoles) of III. The mixture was heated to 40°C and stirred
for two hours. The solvent was removed by rotary evaporation, followed
by vacuum stripping at 80°C to remove the excess 1,5 dibromopentane.
This yielded 1.6 g of I (IoPHBr), a dark viscous liquid which crystal-
lized after two days into a soft white mass. Volhard titrations and
NMR spectroscopy confirmed its structure and purity.

The phenothiazine sulfides were obtained from West Chemical
Company. Their preparation has been described by Nodiff and Cantor.[27]
Their structures, and that of chlorpromazine hydrochloride (CPZ.HCl),
are shown in Fig. 2.

Aldrich Company was the source of phenothiazine (PH), N-methylphenothiazine and 7,7,8,8-tetracyanoquinodimethane (TCNQ). Iodine,
obtained from Matheson, Coleman and Bell Company, was resublimed.
CPZ.HCl was purchased from Sigma Company.

The ionenes, Io$_{x,y}$Br (see Fig. 1) were prepared[28] by the Menschutkin reaction in which α,ω dibromoalkanes are reacted with NN-tetramethyl
alkanes in DMF/methanol to yield water-soluble linear polyelectrolytes.

The magnesium and zinc mesophenyltetrabenzoporphyrins, Fig. 3, were
synthesized and purified by the method described by Reimer and Rose.[29]

A homogeneous mixture of benzylidenephthalimide (25 g, 0.113 mole) and
zinc acetate (10 g, 0.045 mole) was placed in a 100 ml round-bottom
flask. The powdered mixture was heated in a template reactor for one
hour at 350°C under a positive flow of nitrogen. The resultant dark
colored solid material was powdered, extracted with tetrahydrofuran and
filtered through alumina covered with kieselguhr. Purification was
accomplished by repeated chromatography with tetrahydrofuran/hexane
as eluent and alumina as stationary phase. After vacuum drying the
material was recrystallized from a chloroform-hexane mixture. The
purity of the compounds was determined[30] by microanalysis and semi-
quantitative Fast Atom Bombardment mass spectrometry.

Each of the metalloporphyrins synthesized was found to be mixtures
of mesomonophenyl, mesodiphenyl, and mesotriphenyl tetrabenzporphyrin
metallo complexes. As a result the average molecular weights used for
stoichiometric calculations were: MgMTBP, 633 a.m.u.; ZnMTBP, 731 a.m.u.

Iron phthalocyanine was prepared by the method of Barrett et al.[31];
and poly(4-vinyl)pyridine was obtained from Reilly Tar and Chemical Co.

Alternating current titrations were performed with a Wayne Kerr
B221 Autobalance bridge. For these titrations all donors and acceptors
were dissolved in methylene chloride and made up to $10^{-4}$ M. The teflon
conductivity cell was provided with 1 cm square gold electrodes. Adduct
stoichiometries in solution were taken to be the guide for those of the
solid adducts prepared by dry grinding. The stoichiometry for the
latter compounds was checked by solid titration and found to confirm
the ac data.

Pellets for dark conductivity and activation energy examinations
were compressed in a KBr die at 12,000 lbs/inch² with a Carver hydrau-
lic press, although pressure of 3,500 lbs/inch² was generally found to
be adequate for reaching maximum conductivity. Graphite electrodes for
the pellets were cut from a 0.05 mm thick sheet. Dark dc resistivities
and thermal activation energies were measured in a temperature controlled Faraday cage with a Keithley 601C electrometer.

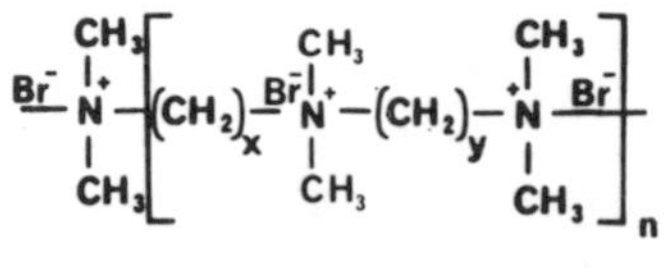

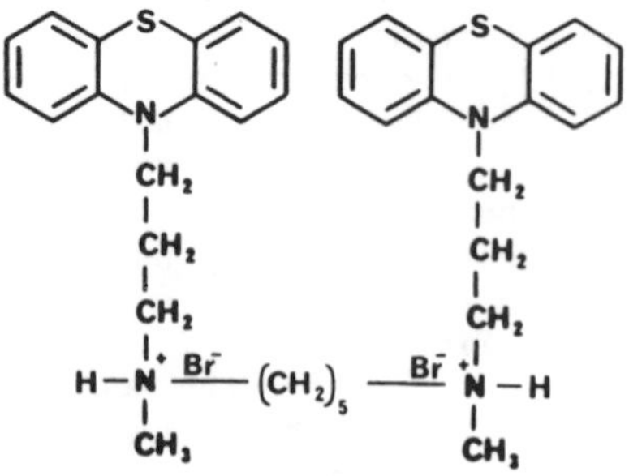

Figure 1. Ionene polymers, $Io_{x,y}X$, where x,y are integers, and X in
this case is Br⁻. IoPHBr is the pentyl dimer of 10-(3-methylamino-
propyl)phenothiazine.

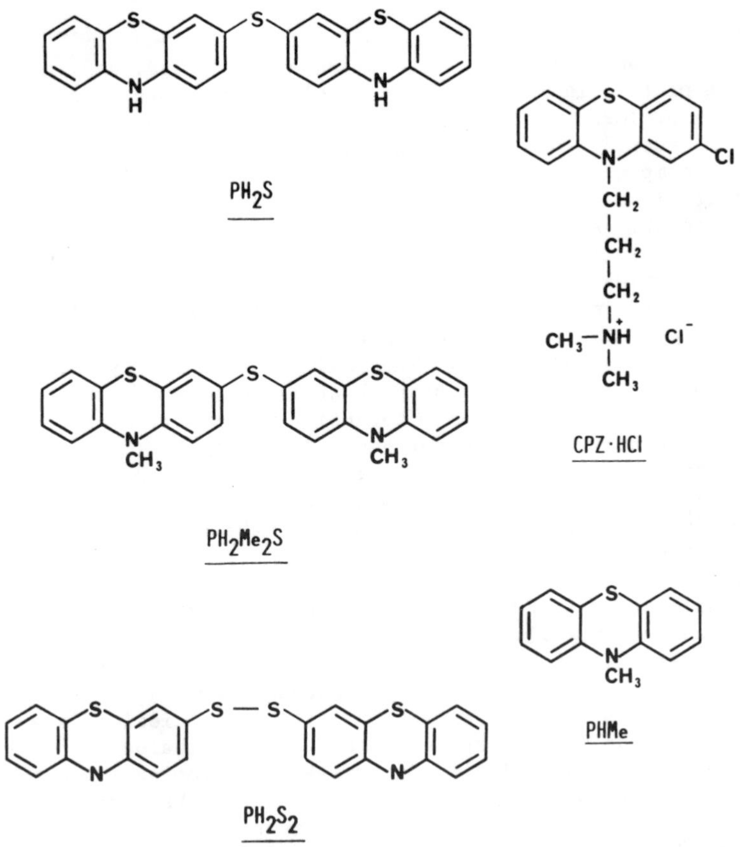

PH$_2$S

PH$_2$Me$_2$S

CPZ·HCl

PHMe

PH$_2$S$_2$

Figure 2. PH$_2$S is phenothiazinylmonosulfide, PH$_2$S$_2$ is phenothiazinyldisulfide, PH$_2$Me$_2$S is N-methylphenothiazinylsulfide. CPZ.HCl is chlorpromazine hydrochloride. PHMe is N-methylphenothiazine.

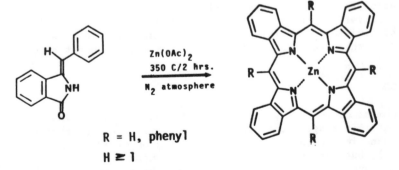

R = H, phenyl

H ≥ 1

Figure 3. One-step template synthesis of the metallo-mesophenyltetrabenzporphyrins.

RESULTS.

Alternating current data are given in Table I and Figs. 4-9. All the ac
curves were plotted in mole fraction of electron donors. The initial
conductances of IoPHBr and TCNQ were so mismatched that a normalized
plot was in order.[32] This plot, Fig. 6, led to an unambiguous determination of the complex stoichiometry as 1:2. In general the forward and
reverse titrations coincided quite well in terms of peak height, see
Figs. 5, 8 and 9. A notable exception was CPZ.HCl versus $I_2$, Fig. 7.
The mismatch, due to micelle formation,[13,33] does not affect the stoichiometry determination. Problems were encountered with the reverse
titrations of PVP into $I_2$ solution, Fig. 4, and that of ZnMTBP into TCNQ
solution. These complexes had a tendency to precipitate from solutions
rich in acceptor concentration. The maximum conductivity information
refers to the the conductivity at peak stoichiometry for the complexes
involved. Almost no electrical interaction was observed between MgMTBP
and PVP, and relatively little between PVP and $I_2$, Fig. 4.

In methylene chloride the IoPHBr:$I_2$ complex was red pink, similar
to CPZ.HCl:$I_2$ while the IoHBr:TCNQ was pink brown. The metal porphyrin-
TCNQ complexes precipitated on standing. All the ac curves were plotted
in mole fraction of electron donors.

Resistivities and thermal activation energies for polyvinylpyridine
(PVP), phthalocyanines (PHTH) and mesophenyltetrabenzporphyrins (MTBP)
are displayed in Table II. Table III exhibits the resistivities and
activation energies for the phenothiazines (PH), chlorpromazine hydrochloride (CPZ.HCl), and ionenes (Io$_x$Br and IoPHBr).

The $I_2$ solid state complexes of all the phenothiazines are reddish
brown, those of the metal porphyrins and the Fe phthalocyanine are deep
blue, or blue green and brittle. The MgMTBP:PVP complex is dark green.
The iodine complexes of all the compounds, except PVP, are very stable,
i.e., readily retain $I_2$. The order of decreasing stability is
phenothiazines, metalloporphyrins, metallophthalocyanine, ionenes,
polyvinylpyridine.

With PVP the brittleness of the porphyrin pellets is ameliorated.
The $I_2$ complexes of CPZ.HCl and IoPHBr tend to be tars, the consistency
of which improved with the addition of PVP; increasing the PVP molar
concentration to beyond 2 however, yielded elastic products.

The activation energy of the complexes dropped considerably, as did
their resistivities. The ternary metallo-organic complexes with PVP
exhibited further drops in resistivity at 1:1 molar concentration.


DISCUSSION

The attachment of pendant phenothiazine groups onto an ionene type chain
has not, at this stage, much improved the resistivity compared with
other ionenes in general which are in the region of $10^5$-$10^8$ ohm cm
(Table III), but has dropped the activation energy markedly. See also
Ref 34. Compared to the resistivity of pure phenothiazines or the model
compound chlorpromazine hydrochloride, however, a dramatic improvement is
observed, i.e., a drop from $10^{12-15}$ to $10^5$ ohm cm. The iodine complex

Table I - Alternating Current Data of Complexes in Methylene Chloride

| Adduct | Stoichiometry | Maximum Conductivity (Micromho) |
|--------|---------------|----------------------------------|
| IoPHBr:$I_2$ | 1:1 | 4.60 |
| IoPHBr:TCNQ | 1:2 | 0.625 |
| CPZ.HCl:$I_2$ | 1:1 | 0.485 |
| PVP:$I_2$ | 1:1 | 0.068 |
| ZnMTBP:$I_2$ | 1:1 | 2.85 |
| MeMTBP:$I_2$ | 1:1 | 5.25 |
| MgMTBP:PVP | - | - |
| ZnMTBP:TCNQ | 4:1 | 0.333 |
| MgMTBP:TCNQ | 4:1 | 0.525 |

Table II - Dark Conductivity and Thermal Activation Energy (E/2kT) Data for Polypyridine, Phthalocyanines and Porphyrines and Their Adducts

| Adduct/Mixture | Molar Ratio | Resistivity (Ohmcm) | Temperature ($^oC$) | Activation Energy (eV) |
|----------------|-------------|---------------------|---------------------|------------------------|
| PVP:$I_2$ | 1:1 | $8.8 \times 10^4$ | 22 | - |
| ZnMTBP:$I_2$ | 1:1 | $5.4 \times 10^4$ | 23 | 0.61 |
| ZnMTBP:$I_2$:PVP | 1:1:1 | $4.6 \times 10^4$ | 23 | 0.43 |
| ZnMTBP:$I_2$:PVP | 1:1:2 | $5.5 \times 10^4$ | 21 | - |
| MgMTBP | - | $9.9 \times 10^8$ | 25.5 | 0.86 |
| MgMTBP:PVP | 1:1 | $9.2 \times 10^{10}$ | 25.5 | 1.5 |
| MgMTBP:$I_2$ | 1:1 | $2.6 \times 10^4$ | 25 | 0.61 |
| MgMTBP:$I_2$:PVP | 1:1:0.5 | $1.1 \times 10^4$ | 21 | - |
| MgMTBP:$I_2$:PVP | 1:1:1 | $1.4 \times 10^3$ | 21 | 0.39 |
| MgMTBP:$I_2$:PVP | 1:1:2 | $6.8 \times 10^3$ | 21 | - |
| MgMTBP:$I_2$:PVP | 1:2:1 | $6.5 \times 10^4$ | 22 | - |
| FePHTH:$I_2$ | 1:1 | $3.3 \times 10^4$ | 21 | - |
| FePHTH:$I_2$:PVP | 1:1:1 | $1.1 \times 10^3$ | 21 | - |
| FePHTH:$I_2$PVP | 1:1:3 | $4.5 \times 10^3$ | 22 | - |
| FePHTH:$I_2$:PVP | 1:1:10 | $3.3 \times 10^5$ | 21 | - |

Table III - Dark Conductivity and Thermal Activation Energy
(E/2kT) Data for Ionenes, Phenothiazines and Their Adducts

| Adduct/Mixture | Molar Ratio | Resistivity (Ohmcm) | Temperature ($^oC$) | Activation Energy (eV) |
|---|---|---|---|---|
| IoPHBr | - | $7.6 \times 10^5$ | 21 | 0.62 |
| IoPHBr:$I_2$ | 1:1 | $4.1 \times 10$ | 22 | - |
| IoPHBr:$I_2$:PVP | 1:1:1 | $7.4 \times 10$ | 22 | - |
| IoPHBr:$I_2$:PVP | 1:1:5 | $1.4 \times 10^5$ | 22 | - |
| $PH_2S$:$I_2$ | 1:3 | $2.6 \times 10$ | 23 | - |
| $PH_2S$:$I_2$:PVP | 1:3:1 | $2.4 \times 10^2$ | 22 | - |
| $PH_2Me_2S$:$I_2$ | 1:3 | $3.3 \times 10^3$ | 23 | - |
| $PH_2Me_2S$:$I_2$:PVP | 1:3:1 | $1.2 \times 10^6$ | 22 | - |
| $PH_2S_2$:$I_2$ | 1:3 | 5.2 | 24 | - |
| PHMe:$I_2$ | 2.3 | 6.8 | 24 | - |
| PH | - | $10^{12}$ | 22 | 1.5 |
| PH:$I_2$ | 2:3 | $3.8 \times 10^2$ | 22 | 0.38 |
| PH:$I_2$:PVP | 2:3:1 | $5.7 \times 10^2$ | 22 | - |
| $Io_{6,4}Br$ | - | $3.1 \times 10^5$ | 21.3 | 0.74 |
| $Io_{6,10}Br$ | - | $1.7 \times 10^6$ | 24 | 1.2 |
| $Io_{6,12}Br$ | - | $6.6 \times 10^5$ | 24.5 | 1.1 |
| CPZ.HCl | - | $10^{15}$ | 23 | 2.4 |
| CPZ.HCl:$I_2$ | 1:1 | $10^5$ | 23 | 2.0 |

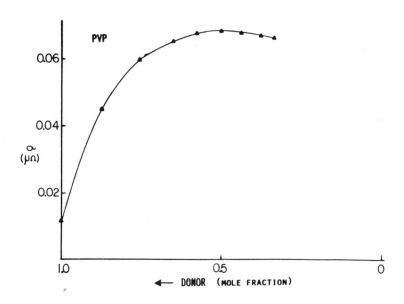

Figure 4.  ac Titration of PVP with $I_2$.

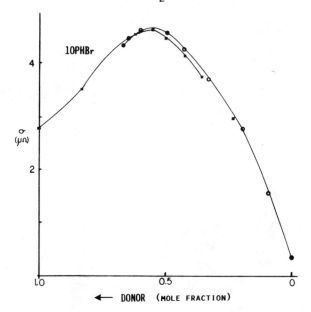

Figure 5.  ac Titration of IoPHBr with $I_2$.

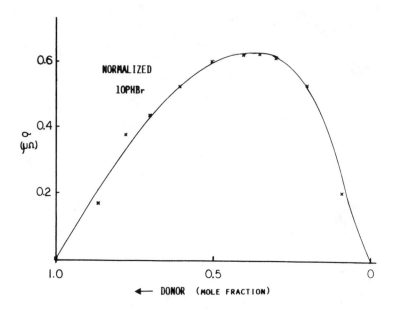

Figure 6.   ac Titration of IoPHBr with TCNQ.

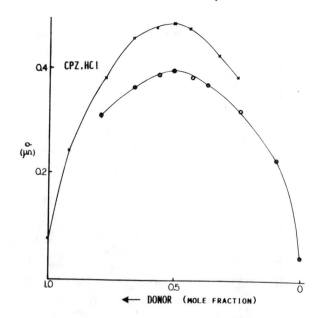

Figure 7.   ac Titration of CPZ.HCl with $I_2$.

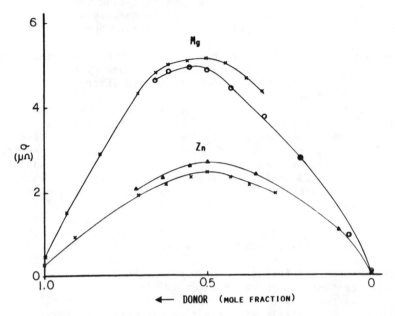

Figure 8.  ac Titrations of MgMTBP and ZnMTBP with $I_2$.

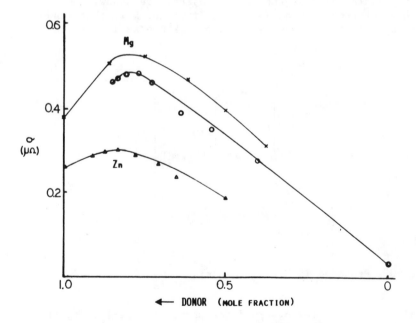

Figure 9.  ac Titrations of MgMTBP and ZnMTBP with TCNQ.

of IoPHBr has provided the lowest resistivity value for an ionene-iodine complex so far determined. This value is in keeping with those observed for the $I_2$ complexes of the pure phenothiazines. See Table III and Ref. 34.

The electrical conductivity of iodine and TCNQ complexes of a number of types of ionene polymers including those with aromatic rings[35] incorporated into the ionene backbone have been extensively studied. The resistivity of aliphatic saturated as well as aromatic ionenes is greatly dependent on the presence of neutral TCNQ. The ionene-TCNQ complexes containing aromatic rings have lower resistivities than TCNQ complexes of completely saturated polymers, which are also less stable.[36]

The electrical properties of polymeric TCNQ complexes seem to depend mainly on the ratio of paramagnetic $TCNQ^-$ to neutral TCNQ and not to be influenced significantly by the nature of the backbone, which has however, an important bearing on their stability. The $I_2$ complexes of all[37] the simple aliphatic ionene salts are quite high, $10^6$-$10^8$ ohm cm.

In the IoPHBr complex the TCNQ molecule may be acting at either the thiazine ring or at the ionic portion of the ionenes. The stoichiometry[38] rules out neither. TCNQ is known to complex 1:1 with phenothiazine,[34] chlorpromazine hydrochloride, and ionenes[37] and nothing extraordinary was observed for its complex with IoPHBr.

A different situation obtains for the interaction of $I_2$ with IoPHBr. It is now well established[13,39] that the interaction of $I_2$ with thiazines occurs primarily with the ring nitrogen. In the IoPHBr dimer the ratio of the phenothiazine groups to $I_2$ is 2:1. This unusual* stoichiometry indicates that the iodine molecule is sandwiched between two phenothiazine groups. See Fig. 10. It is unlikely that conduction in this complex is predominantly via the backbone as is the case generally

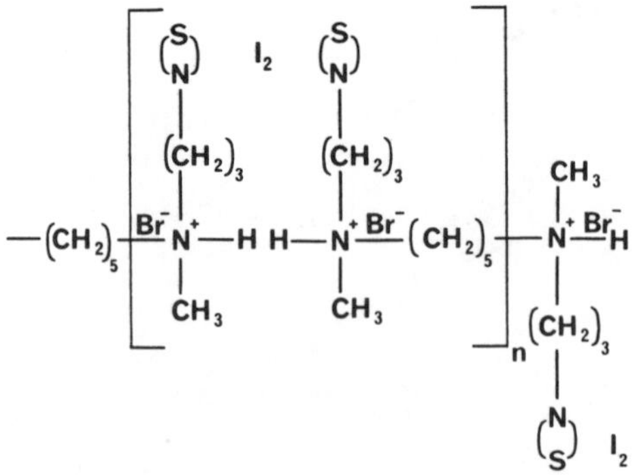

Figure 10. Proposed IoPHBr:$I_2$ complex.

* Usual stoichiometries observed for phenothiazine:$I_2$ complexes are 1:1, 1:2 and 2:2.[32,34,38,40]

with ionene polymers.[34]

In view of the encouraging low activation energy of IoPHBr and the low resistivity of the complexes, polymerization of IoPHBr and similar compounds may thus well lead to sought for improved conductivity characteristics. Summers and Litt,[41] investigating charge transfer complexes of N-methylphenothiazine and related polymeric N donors with acceptors like TCNQ and tetracyanotethylene, found that the electronic conductivity was higher in the polymers than in the model complexes and increased with increasing crystallinity. Thermal activation energies for the complexes ranged from the 0.5-1.8 eV and were roughly the same for the model and the polymer complexes.[41]

The addition of PVP to the $I_2$ complexes of the IoPHBr dimer and CPZ.HCl improves their mechanical properties considerably. However, PVP acts as a diluent causing resistivities to increase in all the phenothiazine:$I_2$ complexes.

Quite a different situation occurs in the metallophthalocyanine and porphyrin complexes of $I_2$. Here PVP is an active participant. The resistivity in the solid state is depressed at the molar ratios 1:1:1 1:1:1 (see Table II) by an order of magnitude in the cases of FePHTH and MgMTBP. The latter complex shows a dramatic further drop in activation energy. This effect is thought to be due to a ternary interaction, rather than an interaction between the PVP:$I_2$ (1:1) complex and the 1:1 metallo-organic:$I_2$ complex as shown by the increase in resistivity of MgMTBP:$I_2$:PVP = 1:2:1.

The stoichiometry of the TCNQ complexes of MgMTBP and ZnMTBP is 1:4, indicating that the porphyrins are oriented with the metal core towards the N of the electron acceptor. This interaction is weak, but the interaction between the metallo-porphyrins and PVP is, by comparison, non-existent. See Table I. It has been observed that zinc tetraphenylporphyrin complexes poorly to PVP compared to pyridine.[42] In fact, mixing PVP with MgMTBP in the solid state raises the resistivity and the activation energy of MgMTBP markedly. See Table II. The 1:1:1 interaction products of the metalloporphyrins, iodine and polyvinylpyridine most likely involve a ternary arrangement in which the N of PVP couples to the $I_2$ in the metal MTBP:$I_2$ stacks. See Fig. 11. EPR data demonstrated[43] the delocalization of $\pi$ electrons in the direction of this kind of ligand stack. EPR data[44] showed that preferential oxidation of the ligand takes place inferring that the complexed metals do not play an important role in the conduction process of iodinated metalloporphyrins and metallophthalocyanines. The charge carriers are associated with delocalized $\pi$-orbitals on the macrocycle. An exception is Ni-tetrabenzporphyrin in which charge is transferred both via ligands and the metal.[19]

Freeman[44] confirmed that metallo-organics can accept or donate electrons without affecting the coordination number of the metal in poplar plastocyanin, which has a free Cu center. This center is embedded deep in the protein and has no contact with the external environment.

It is probable that the $(MgMTBP:I_2)_n$ stack is poorly aligned on the axis through the Mg atoms, especially in view of the fact that the meso compounds were mixtures (see Materials and Methods), and the PVP, complexing with the $I_2$ chain, causes this chain to marshall the porphyrin rings into a more orderly stack. Alternatively the porphyrin

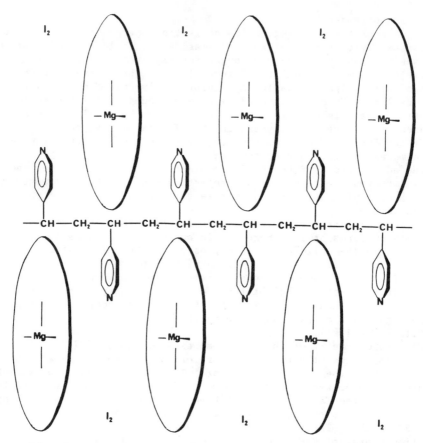

Figure 11.  Proposed structure of ternary complex of MgMTBP:$I_2$:PVP.

Figure 12.  Representation of conduction pathway of stacked metallo-
porhpyrins and phthalocyanines.  The top schematic illustrates the
conduction pathway for the nickel tetrabenzporphyrin.  After Hoffman
and Ibers.

stack may impose rigidity on the PVP polymer.

Hadek and Ulbert[46] showed that for charge transfer complexes the electron conduction was governed by the acceptor component, and the hole conductance by the donor component. Hadek[47] concluded that the electron conductivity in, say, the N,N'-diphenyl-p-phenylenediamine:$I_2$ complex at a higher temperature was dominated by the polyiodide chains, while the lower temperature hole conductivity was governed by the relatively closely spaced donor centers. A similar situation might obtain here. However, this needs to be proven.

The structures of charge transfer complexes relative to their electrical properties have received considerable attention. Tamamura and Yamane[48] were able to show that charge transfer interactions between cations and anions affected crystal structure to a lesser degree than hydrogen bonding or electrostatic attraction. Charge transfer stabilization energies of $\pi - \pi^*$ complexes are unlikely to be as large as those of London dispersion, charge-dipole, dipole-induced dipole or hydrogen bonding interactions, so that charge transfer interactions should have an effect on donor-acceptor overlap only when these larger forces show little sensitivity to orientation. Some exceptions, say, should be expected where such stabilization would be maximized by a center-on-center orientation[49]. It is of interest to note that the most frequently found interplanar distance for solid charge transfer complexes is 3.3 Å. In charge transfer complexes which have segregated stacks of organic donor and acceptor molecules the charges are partially localized and may vary from site to site down the stack, forming a modulated charge density, or a charge density wave. Fig. 12 shows the schematic of stacked porphyrins and phthalocyanines[50]. In such a crystal, i.e., say tetrathiafulvalene-TCNQ, attractive interactions exist in some directions and repulsive interactions in others.

The stability of ternary systems depends greatly on aromatic stacking interactions[6]. Such "indirect charge transfer" interaction in multi-layered organics has been discussed by Staab[51]. Similar arrangements have been found[52] to exist in metallocenes of the general formula $M(C_nH_n)_2$ where M stands for a metal valency of 2,3 or higher. These sandwich complexes are said[52] to form a "skyscraper-like" arrangement.

The entropy contributions to the formation, and stability, of ternary complexes must be quite complicated and do not appear to have been studied to any extent. Copolymers of di-methylamino-styrol as donors and e.g., chloranil as acceptor, have been studied[18] in detail: by complexation the segmental mobility of the macromolecules is reduced so that the entropy likewise is lowered causing a drop in the enthalpy of formation. It is highly likely that in the proposed metallo-organic: $I_2$:PVP model an entropy advantage is obtained by improved stacking of the metallo-organic:porphyrin:$I_2$ complexes, the stack imposing rigidity on the PVP, or on a mutually reinforcing interaction leading to the observed improved resistivities and activation energies. It is appreciated that this model is but one of several possibilities. However, its appeal is that it fits all the data best.

## ACKNOWLEDGEMENT

We wish to thank Dr. Cantor for his generous gift of phenothiazine
sulfides.

## REFERENCES

1.  S.F. Fischer et al., Bull. Amer. Phys. Soc., 24, 346 (1979).
2.  N. Karl, Pr. Nauk. Inst. Chem. Org. Fiz. Politech. Wroclaw, 16, 43
    (1978).
3.  R. J. P. Williams, Current Topics in Bioenergetics, 3, 79 (1969).
4.  J. Mordclewski and W. E. Geiger, J. Am. Chem. Soc., 100, 7429 (1978).
5.  A. Van Heuvelen, J. Biol. Phys., 1, 215 (1973).
6.  H. Siegel and C. F. Naumann, J. Am. Chem. Soc., 98, 730 (1976); H.
    Siegel and P. E. Amisler, ibid., 98, 7390 (1976); H. Siegel et al.,
    Inorg. Chem., 16, 790 (1977); E. C. Johnson et al., Canad. J. Chem.,
    56, 1381 (1978); P. R. Mitchell and H. Siegel, J. Amer. Chem. Soc.,
    100, 1564 (1978); Helv. Chim. Acta, 62, 1723 (1979); C. R. Naumann
    and H. Siegel, FEBS Lett., 47, 122 (1974); P. Chaudhuri and H.
    Siegel, J. Amer. Chem. Soc., 97, 3209 (1975); 99, 3242 (1977); H.
    Siegel et al., Europ. J. Biochem., 41, 290 (1974); G. Cilento,
    Quart. Revs. Biophys., 6, 488 (1973); Y. Fukuda, P. R. Mitchell and
    H. Siegel, Helv. Chim. Acta, 61, 638 (1978); H. Siegel et al.,
    Biochim. Biophys. Acta, 148, 655 (1967); H. Siegel and P. E. Amsler,
    ibid., 98, 7390, (1976); A. M. Golub, Russ. Chem. Revs, 45, 479
    (1976) R. Clement et al., J. Chem. Soc. Chem. Commun., 654 (1974);
    A. N. Nesmeyanov et al., Dokl. Nauk, SSSR, 221, 229 (1975); P.
    Pasman et al., Chem. Phys. Lett., 59, 398 (1978); L. M. Titvinenko
    et al., Russ. Chem. Revs, 44, 718 (1975).
7.  H. Bretschneider, Wiss. Z. Tech. Hochschule Karl Marxstadt, 17, 281
    (1975); F. Cramer, Einschlussverbindungen, Springer, Heidelberg
    (1974).
8.  Y. Matsunaga et al., Bull. Chem. Soc. Japan, 47, 2826 (1974).
9.  M. Mohwald, Chem. Phys. Lett., 26, 509 (1974).
10. H. Siegel and D. B. McCormick., J. Am. Chem. Soc., 93, 2041 (1971).
11. V. Hadek, H. Noguchi and A. Rembaum, Macromolecules, 4, 494 (1971).
12. A. T. Florence, in Micellization, Solubilization and Microemulsions
    (K. L. Mittal, ed.) Plenum Press, New York (1977), Vol. 1.
13. P. K. Dea and H. Keyzer in Modern Bioelectrochemistry, (F. Gutmann
    and H. Keyzer, eds.), Plenum Press, New York (1986).
14. T. E. Phillips, B. M. Hoffman, J. Amer. Chem. Soc., 99:23, 7734
    (1977).
15. J. L. Peterson, C. S. Schramm, D. R. Stojakovic, B. M. Hoffman,
    T. J. Marks, J. Amer. Chem. Soc., 99:1, 286 (1977).
16. S. K. Wright, C. J. Schramm, T. E. Phillips, D. M. Scholler,
    B. M. Hoffman, Synthetic Metals, 1, 43 (1979).
17. T. E. Phillips, R. P. Scaringe, B. M. Hoffman, J. A. Ibers, J.
    Amer. Chem. Soc., 102:10, 3435 (1980).
18. J. Martinsen, L. J. Pace, T. E. Phillips, B. M. Hoffman and J. A.
    Ibers, J. Amer. Chem. Soc., 104, 83 (1982).

19. W. B. Euler, J. Martinsen, L. J. Pace, B. M. Hoffman and J. A. Ibers, Mol. Cryst. Liq. Cryst., 81, 949, 231 (1982).
20. B. M. Hoffman and J. A. Ibers, Acc. Chem. Res., 16, 15 (1983).
21. T. Inabe, S. Nakamura, W. Liang and T. J. Marks, J. Amer. Chem. Soc., 107, 7224 (1985).
22. J. Metz and M. Hanack, J. Amer. Chem. Soc., 105, 828 (1983).
23. M. Hanack, Israel J. Chem., 25, 205 (1985).
24. P. G. Wright, P. Stein, J. M. Burke and T. G. Spiro, J. Amer. Chem. Soc., 101, 3531 (1979).
25. G. Geib and H. Keyzer, In press.
26. I. Cozlan et al., J. Heterocyclic Chem., 21, 613 (1984).
27. E. A. Nodiff and A. Cantor, "Phenothiazine Sulfides," U.S. Patent; 4,155,874, May 22, 1979.
28. e.g., H. Noguchi and A. Rembaum, Polymer Letters, 7, 383 (1969).
29. K. G. Reimer and C. B. Rose, In Press. A. S. Nelson and S. P. Gottfried J. Amer. Chem. Soc., 63, 487 (1941).
30. K. G. Reimer, D. Sensharma, J. Wells, In Prep.
31. P. A. Barrett, D. A. Frye and R. P. Linstead, J. Chem. Soc., 1157 (1938).
32. F. Gutmann and H. Keyzer, Electrochim. Acta, 11, 1163 (1966).
33. H. Keyzer and S. Maurer, In Prep.
34. F. Gutmann, H. Keyzer and L. E. Lyons, Organic Semiconductors. Part B, Robert E. Krieger Publishing Co., Malabar, Florida (1983).
35. A. Rembaum, Encyclopedia of Polymer Sci. and Technol., 11, 318 (1968) Wiley, N.Y., 1969; V. Hadek et al., Makromolecules, 4, 494 (1971).
36. J. M. Bruce and J. R. Henson, Polymer, 8, 619 (1967).
37. H. Keyzer, unpublished results.
38. F. Gutmann and H. Keyzer, Electrochim. Acta, 11, 555 (1966).
39. S. Chan, C. M. Gooley and H. Keyzer, Tetrahedron Lett., 13, 1193 (1975).
40. F. Gutmann and H. Keyzer, J. Chem. Phys., 46, 1969 (1967).
41. J. W. Summers and M. H. Litt, J. Polym. Sci., Polym. Chem. Ed., 11, (6), 1379 (1973).
42. D. S. Becker and R. G. Hayes, Inorg. Chem., 22, 3050 (1983).
43. C. S. Schramm, R. P. Scaringe, Dr. R. Stojakovic, B. M. Hoffman, J. A. Ibers and T. J. Marks, J. Am. Chem. Soc., 102, 6702 (1980).
44. L. J. Pace, A. Ulman and J. A. Ibers, Inorg. Chem., 21, 199 (1982).
45. H. C. Freeman, J. Proc. Royal Soc. N.S.W., 112, 60 (1979).
46. V. Hadek and K. Ulbert, Rev. Sci. Instr., 38, 991 (1967).
47. V. Hadek, J. Chem. Phys., 49, 5202 (1968).
48. J. Tamamura and T. Yamane, Bull. Chem. Soc. Japan, 47, (4), 832 (1974).
49. J. C. A. Boeyens and I. H. Herbstein, J. Phys. Chem., 69, 2160 (1965).
50. J. H. Torrance and B. D. Silverman, Phys. Rev. B, 15, (2) 788 (1977).
51. H. A. Staab et al., Angew. Chemie Int. Ed., (Engl.), 16, 801 (1977).
52. H. Werner, Angew. Chem., 89, 1 (1977); Angew. Chem. Int. Ed. (Engl), 16, (1977).
53. D. Braun and H. J. Sterzel, Ber. Bunsen Ges. Phys. Chem., 76, 551 (1972).

# INCREASING THE CONDUCTIVITY OF POLYACETYLENE FILMS BY ELONGATION

N. Theophilou, H. Naarmann[*]
Kunsttoff-Laboratorium BASF
6700 LUNDWIGSHAFEN (F.R.G.)

## Abstract

An acetylene polymerization reaction catalyzed by an aged mixture of $Ti(OC_4H_9)_4/Al(C_2H_5)_3$ in a silicone oil reaction medium yields a homogeneous, defect-free polyacetylene film that can be stretched mechanically by up to 600% corresponding to stretching rates of 7. After we had doped with iodine, the film displayed electrical conductivity of ca. 11 000 S/cm. Both washed (catalyst-free) and unwashed (catalyst-containing) film was stretched.

## Introduction

A decisive step towards the production of polyacetylene was the determination of the direct proportionality between conductivity and crystallinity and the indirect relationship to the $sp^3$ amount (1).

By modifying the polymerization conditions, e.g. by using silicone oil, we can polymerize $HC \equiv CH$ at room temperature to yield a $N-(CH)_x$ polyacetylene of at least the same quality as the $S-(CH)_x$ that is obtained at $-78^oC$. The results are contained in Table 1 (2).

Aging the standard catalyst results in a surprising improvement in the $(CH)_x$ properties (3). The reduction in the number of $sp^3$ orbitais, i.e. the production of a defect-free system, is a great advantage.

Special techniques have been adopted to orientate the $(CH)_x$ to obtain high conductivity values, i.e. values greater than 10 000 S/cm).

Shirakawa (4) developed a method for obtaining highly conducting $(CH)_x$ by polymerization of $CH \equiv CH$ in an oriented liquid crystal matrix. This method is similar to that of Aldissi (5). In our case, we stretched the film mechanically up to 600 % - corresponding to stretching rates of 7.

L. Alcácer (ed.), Conducting Polymers, 65–75.
© 1987 by D. Reidel Publishing Company.

TABLE 1
Properties of the different $(CH)_x$ types

| crystallinity[a] % | conductivity[b] (S/cm) | | sp$^3$ content[c] (rel.%) | surface[d] area (m$^2$/g) | configuration [e] cis content % |
|---|---|---|---|---|---|
| | undoped | doped with $I_2$ | | | |
| S-$(CH)_x$   70 | $10^{-6}$ | 200 | 4 | 300 | 50 |
| N-$(CH)_x$   65 | $10^{-6}$ | 2000 | 0 | 100 | 80 |

a) Philips diffractometer CuK$_\alpha$ radiation; (b) four-probe
measurement; (c) determined by $^{13}$C NMR spectroscopy; (d) BET-method;
(e) determined by IR spectroscopy

Preparation of catalyst and $(CH)_x$ samples

Apparatus

500-ml four-neck flask with thermometer, dropping funnel, magnetic
stirrer, and connection for vacuum and argon.

Reactivity mixture

50 ml Silicone Oil AV 1000 (Wacker)
31 ml triethylaluminium - TEA, $(C_2H_5)_3Al$
41 ml tetrabutoxytitanium $Ti(C_4H_9O)_4$ freshly distilled (Dynamit Nobel)

Procedure

The silicone oil is stirred and degassed for 20 minutes at 0.05 mbar.
The TEA is added in a counter-current stream of argon, and the tetrabutoxytitanium is run in, drop by drop, through the inactivated dropping
funnel over a period of one hour at 38-42 $^o$C. Afterwards, the mixture
is degassed for one hour at room temperature and subsequently stirred
for two hours at 120$^o$C in a weak current of argon. This is followed
by stirring and degassing, the temperature is reduced from 120$^o$C to
room temperature (22$^o$C). The polyacetylene film is produced in a glove
box with the catalyst thus obtained. The evacuated flask with the
reactants is placed in the glove box.

## Production of film in the glove box

An even, homogeneous layer of catalyst is applied onto a flat carrier consisting of glass or any other material, e.g. nylon, polyester film or other polymer-supporting material which is stretchable, e.g. poly-butadiene.

The amount of catalyst required to obtain a $(CH)_x$ film of 30 µm thickness is 7ml. The carrier coated with the catalyst is hermetically sealed within the glove box by means of a hood fitted with a gas inlet valve. First of all, the hood is evacuated; then 600 ml of acetylene is passed into it over a period of 15 minutes (the acetylene is previously purified over BTS-contact and a separate trap for TEA and tetrabutoxytitanium). As it is being admitted, the acetylene polymerizes at the surface of the catalyst on the carrier to $(CH)_x$, which is a black, homogeneous film that can be peeled off from the carrier. The $(CH)_x$ film is extremely porous and contains all the catalyst.

The film is washed under the following conditions:
 3 hours in toluene
16 hours in methanol containing 6 % of hydrochloric acid
two periods of 1-1.5 hours with methanol.

The washed film of 30 µm thickness is dried for about 10 minutes in a current of argon in the glove box followed by 30 minutes at 25°C and 0.05 mbar.

Analysis of the dried film yields the following results:

C:        91.7-92.2  %
H:         7.6- 7.7  %
O:              0.5  %
Al/Ti:         0.01  %

Fig. 1 shows the f.t.i.r. spectra of the $(CH)_x$ film.

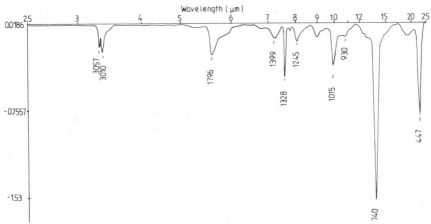

Fig.1:  F.t.i.r.  spectra  of  the  unstretched  $(CH)_x$  film,  cis/trans
        87/13

No bands corresponding to valence-bond resonance of -CH$_3$- or -CH$_2$- could be detected at wave-lengths of 2960, 2910 and 2830 μm in the infrared spectrum. Likewise, the characteristic peaks of sp$^3$ are absent between 30 and 40 ppm in the C$^{13}$ NMR (CxP 200 Bruker instrument). A relationship has been observed in other catalyst systems between the sp$^3$ fraction and the conductivity (2).


## Stretching of the samples

The samples were stretched in the glove box.

These received N-(CH)$_x$ film (Fig. 2) was prepared on the surface of

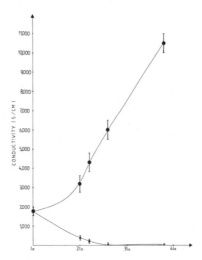

Fig. 2: Conductivity ( σ$_\parallel$ and σ$_\perp$ ) versus elongation of a 30 μm (CH)$_x$
        film, suported on a polybutadiene film - stretched with the
        supporting material.
        The (CH)$_x$/polybutadiene was stretched with the catalyst system.
        After it had been stretched, the (CH)$_x$ film was removed from
        the polybutadiene film, washed in the normal way and doped.

a polybutadiene film, stretched, removed from the supporting film and washed in the usual fashion with toluene, CH$_3$OH/HCl, washed with methanol, dried and then doped with iodine. Another N-(CH)$_x$ film (Fig. 3) which was used for these measurements was prepared without any organic supporting material, washed in the usual way, stretched and then doped. Fig. 4 shows that the highly stretched film has surprisingly better air stability. Fig. 5 is a s.e.m. of a 500 % stretched (unwashed) N-(CH)$_x$ film; stretching was performed in the presence of the catalyst. Fig. 6 is a s.e.m. of a 500 % stretched (washed) N-(CH)$_x$ film; stretching was performed after washing. Fig. 7 is a s.e.m. of a S-(CH)$_x$ film stretched 170 % after washing; S-(CH)$_x$ film that was subjected to higher stretching rates was not flexible.

Comparing Fig. 6 with 7, we can distinguish completely different structures: long fibrils in the case of the stretched N-(CH)$_X$, and the well-known coiled (CH)$_X$structure of S-(CH)$_X$ in Fig. 7.

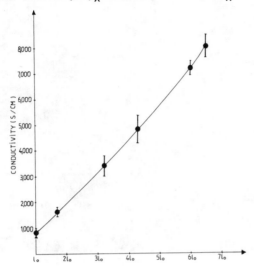

Fig. 3: Conductivity ( $\sigma_{//}$ ) versus elongation of a 50 m (CH)$_X$ film, stretched after washing and then doped.

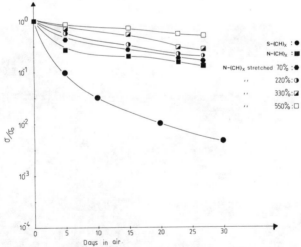

Fig. 4: Normalized conductivity versus days of ambient air exposure for I$_2$-doped, S-(CH)$_X$ film ( ● ), N-(CH)$_X$ film ( ■ ) and stretched 70 % ● 220 % ◑, 330 % ◪ and 550 % ▢ . These films were prepared by washing, stretching and, finally, doping.

Fig. 5: S.e.m. of a N-(CH)$_x$ film, stretched 500 %. Stretching was performed with the catalyst-containing film. After it had been stretched, the film was washed in the normal manner.

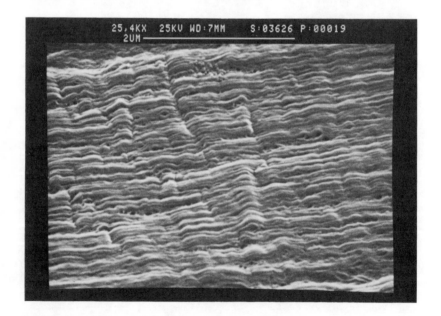

Fig. 6: S.e.m. of a N-(CH)$_x$ film, stretched 500 %. This film was stretched after the washing stage.

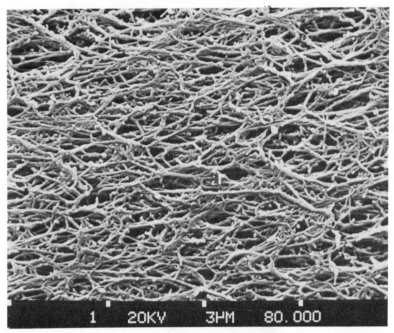

Fig. 7: S.e.m. of a S-(CH)$_x$ film, stretched 170 %. This film was
stretched after the washing stage.

## Doping

The polyacetylene film is doped for one hour in a saturated solution
of iodine in carbon tetrachloride (2.65g of iodine in 100 ml of carbon
tetrachloride at 25 $^oC$). Afterwards, it is washed three times with
carbon tetrachloride for periods of two minutes each, and then dried
for 20 minutes in a current of argon, followed by one hour in 0.1
mbar vacuum at 30 $^oC$.

The mass fraction of iodine in the film after doping is 70 %.

The specimens are sensitive to ligth and add on $CCl_3$ as well as iodine
($CCl_4$ catalyzed $\rightarrow$ $Cl_3$ + .Cl). The reaction mechanism is complicated
and proceeds via a radical chain reaction combined with a hydrogen
abstraction at the (CH) (4).

## Measurement

Four-point measurement

Conductivity:        unoriented
                     $\sim$2000 S/cm

Similarly, transparent (CH)$_x$ film can be made that has conductivity
values greater than 5000 S/cm. The polyacetylene is produced on a

plastic film and stretched jointly with the supporting material.
Afterwards, it is made to form a complex with, e.g. iodine, under
standard conditions.

## Transparent films

By using HDPE (90 μm thick film) as the supporting material for the
preparation of $(CH)_x$, we obtain transparent, highly conducting film.
Fig. 8 shows the f.t.i.r. spectra of the supporting HDPE (a) and the
combination of $(CH)_x$ and HDPE (b).

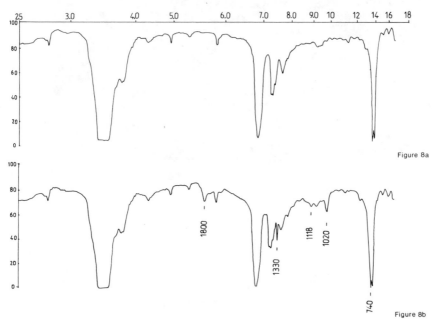

Figure 8a

Figure 8b

Fig. 8: F.t.i.r. spectra (Bruker IFS 85) of a 90 μm thick HDPE
transparent film (a) and of a 2 μm thin (transparent) $N-(CH)_x$
film supported on a HDPE film (90 μm) (b).

Frequencies
of $(CH)_x$:        $1800$ cm$^{-1}$ = combination of the C-C-C deformation
                                with the vibration of C-H in the cis
                                $(CH)_x$ plane
            $1330$ cm$^{-1}$ = cis C-H in-plane
            $1118$ cm$^{-1}$ = valence band of C-C cis
            $1020$ cm$^{-1}$ = trans C-H out-of-plane
             $740$ cm$^{-1}$ = overlapping of cis C-H out-of-plane on
                                absorption of HDPE film

Fig. 9 shows different samples of the joint systems of $N-(CH)_x$ and
HDPE film were stretched to various lengths, washed and doped.

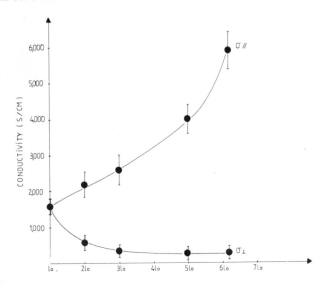

Fig. 9: Conductivity ( $\sigma_{\parallel}$ and $\sigma_{\perp}$ ) versus elongation of 0.3 $\mu$m transparent N-(CH)$_x$ film, supported on a HDPE foil and stretched with the supporting material. The N-(CH)$_x$/HDPE film was stretched before being washed. After it had been washed the film was doped with I$_2$.

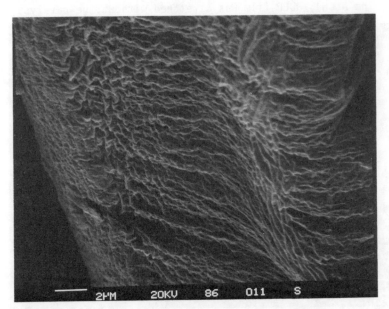

Fig. 10: S.e.m. of oriented, long fibrils of a transparent N-(CH)$_x$, supported on a HDPE film, stretched 500 %

Fig. 10 shows fibrils of transparent $N-(CH)_X$ prepared by the method above. Fig. 11 shows the conductivity of different transparent N-$(CH)_X$ samples with variuos elongation rates.

Stretching was performed after the film had been washed. Fig. 12 is a s.e.m. of a transparent $N-(CH)_X$ film. In this case, stretching was carried out after washing.

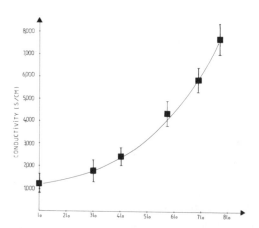

Fig. 11: Conductivity ( $\sigma$ ) versus elongation of a 0.6 m transparent $N-(CH)_X$ film, supported on a HDEP foil, washed in the normal way, stretched with the supporting material and then doped. The $N(CH)_X$/HDEP film was stretched without the catalyst system.

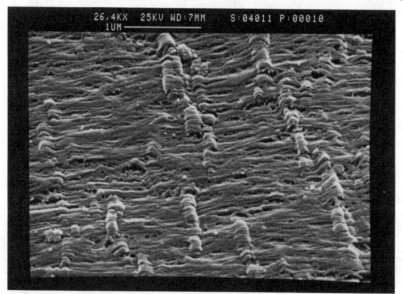

Fig. 12: S.e.m. of oriented, long fibrils of a transparent $N-(CH)_X$ film, supported on a HDPE film (stretching 670 % after washing).

## ACKNOWLEDGEMENTS

I am indebted to H. Heckmann for REM, H. Haberkorn for X-ray, P. Simak for IR and R. Voelkel for $C^{13}$ measurements.

Thank to G. Köhler for $(CH)_x$ preparations.

This work is part of project No. 03-Cl-340 sponsored by the BMFT. The author thanks the Budesministerium für Forschung und Technologie for supporting this work.

## REFERENCES

(1) Structure and conductivity of $(CH)_x$, H. Haberkorn et al., Synthetic Metals, 5 (1982) 51

(2) $(CH)_x$ synthesis in silicone oil at room temperature, EP 88301 Mar.5, 1982/Feb. 25, 1983, BASF Germany

(3) Synthetic metals, H. Harberkorn et al., publication in preparation

(4) H. Shirakawa et al., Synthetic Metals, 14 (1986) 173 and 199

(5) M. Aldissi, J. Polym. Sci. Polym. Lett. Ed. 23 (1985) 167.

# ANISOTROPIC PROPERTIES OF ORIENTED DURHAM ROUTE POLYACETYLENE

P. D. Townsend
Cavendish Laboratory
Madingley Road
Cambridge CB3 0HE,
England

ABSTRACT. Polyacetylene prepared by the Durham precursor route can be obtained in a highly-oriented and crystalline form if a free-standing precursor film is stretched during the transformation reaction. Films produced in this way show similar "long-chain" characteristics to Shirakawa polyacetylene in which it is considered that spin and charge defects have considerable mobility along the polyene chains. These films, being non-fibrous offer the possibility of studying the intrinsic anisotropic properties of the material. Here, the results of polarisation-dependent measurements of optical absorption, reflectivity, photoinduced absorption (PA) and photoconductivity (PC) are discussed. These measurements demonstrate that optical absorption of band-gap light polarised with E vector perpendicular to the chains is due to inter-chain excitation from $\pi$ to $\pi^*$ states. It is found that the long-lived photoexcited states which give rise to PA and PC are preferentially excited by these inter-chain transitions.

## 1.    INTRODUCTION

The last decade has seen a remarkable growth of interest in conjugated polymers as new electronically active materials with potential commercial applications [1]. However, a major problem for workers in the field has been the lack of processibility and morphological control which is inherent to many of the materials synthesised to date. Consequently, the effects of the (often complex) polymer morphology on the observed electronic properties has been only poorly characterised. One way in which this problem can be overcome is by the use of "precursor" routes in which the processibility of a conventional polymer is combined with the desirable electronic properties of a conjugated polymer by the use of an intermediate or "precursor" polymer [2]. The usual scheme is to find a non-conjugated, solution processible polymer which can be easily transformed by for example an elimination reaction, to give the conjugated product. This approach has important advantages over other synthetic routes. Among these is the possibility of varying the morphology of the conjugated polymer by altering the conditions of the transformation step. Differences in the electronic properties of the various morphological forms can then be used to give information on the role of structure etc. in determining these properties. As will be discussed, of particular interest is the ability to produce highly-oriented films which allow the intrinsic anistropic properties of the polymer to be studied. In this contribution I review some of the work that has been carried out over the last two-years in our laboratory and elsewhere on a highly-oriented form of polyacetylene produced via the

*L. Alcácer (ed.), Conducting Polymers, 77–87.*
© *1987 by D. Reidel Publishing Company.*

Durham precursor route.

## 2.   FILM PREPARATION AND CHARACTERISATION

The starting point for the production of oriented films is the Durham precursor route (figure 1) which was developed by Feast and coworkers [3]. This route utilises a soluble, non-conjugated precursor polymer, polymer B, which can be converted to a fully dense form of polyacetylene by the thermal elimination of hexafluoroorthoxylene [4-6]. This conversion process occurs in three stages [6,15,16]. The initial elimination of aromatic units from the polymer (transformation) is followed by evaporation of the volatile product from the polymer and, finally, isomerisation of the cis-rich material formed after transformation to the trans-isomer. The kinetics of these reactions have been extensively studied using DSC, i.r. spectroscopy and weight-loss measurements [6].

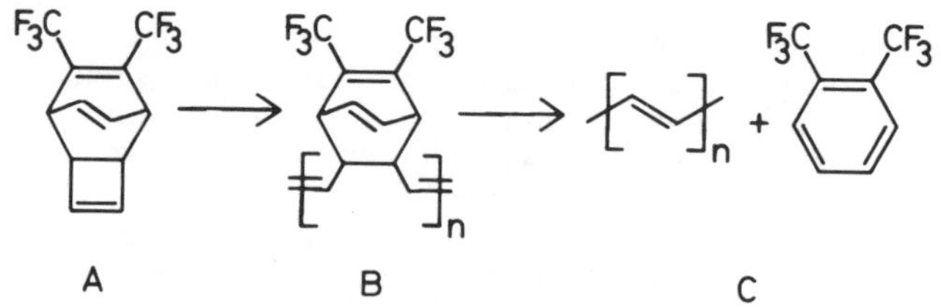

Figure 1. The Durham route to polyacetylene.

As the precursor polymer is non-conjugated it is expected to be randomly coiled and entangled in solution and this disorder is necessarily retained in precursor films cast by solvent evaporation onto substrates. Thermal conversion then gives a highly disordered form of polyacetylene since, in spite of the now conjugated polymer chains which would prefer to be straight, there is no opportunity for the necessary rearrangements of the chains within the solid film. Hence, Durham polyacetylene in this morphology shows only a very low degree of crystallinity and it is considered that the straight chain sequences in the material are only very short, no longer than 20-30 carbon atoms [6]. This is despite the fact that the full polymer chains contain between $6 \times 10^3$ and $1.5 \times 10^4$ carbon atoms [7]. The transport and optical properties of these films are significantly different from those of Shirakawa polyacetylene, reflecting the structural differences of the two materials [5, 8].

In contrast, if a uniaxial stress is applied to a free-standing film of the precursor polymer during the thermal conversion reaction, the film can be stretch-aligned [9] and the resulting polyacetylene shows a very high degree of local ordering and orientation [10-12]. This process was carried out under dynamic vacuum in a temperature controlled glass jacket. At a temperature of 40°C and applied stress of 500-600 N/cm² (for the film, in the precursor form) it is possible to obtain stretch-ratios ($l/l_o$) of up to 20. At this temperature the transformation reaction takes a few hours which gives a convenient time scale for stretching. Films were subsequently isomerised at 80-100°C.

Polyacetylene films produced in this way are well-aligned and show a high degree of crystallinity as evidenced by the x-ray photograph shown in figure 2. This diffraction pattern was obtained from a film of stretch ratio $l/l_o = 11$ [17]. Line profile analysis of the various reflections shows that the polyacetylene films are composed of uniaxially oriented paracrystallites with dimensions of 50Å perpendicular to, and 80Å parallel to the chain

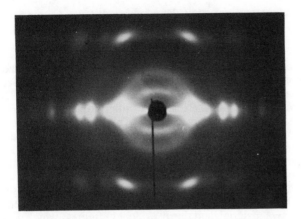

Figure 2. X-ray diffraction photograph of oriented Durham polyacetylene.

direction [10]. Similar values have been obtained by Kahlert et al [18] who have also been able to determine the full three-dimensional structure of the crystallites. Useful information on the degree of disorder in the material can be obtained from the extent of arcing of the equatorial reflections [17]. The intensity distribution along the arc gives a HWHM chain orientation angle to the stretch direction which decreases towards a limiting value of about $3°$ as the stretch ratio increases to $l/l_o > 12$ [17].

## 3. ELECTRONIC PROPERTIES OF STRETCHED AND UNSTRETCHED FILMS

### 3.1. Transport and Magnetic Properties

The transport and magnetic properties of the Durham materials show a transition from behaviour associated with charge or spin defects well localised in the unstretched films to behaviour similar to that found in Shirakawa polyacetylene in which it is considered that the defects have a high mobility along the polyene chains [19, 20]. Thus, unstretched films exhibit a broad Gaussian ESR line with a low temperature width of 10 Gauss which narrows slightly to around 8 Gauss at room temperature. This observation is indicative of immobile defects trapped on short straight-chain or conjugation sequences which are no longer than the soliton width (some 15-20 carbon bonds) [6,21,22]. In contrast, stretched films show considerably narrowed and anisotropic lines (at room temperature with B parallel to the chains $\Delta H = 3$ Gauss, with B perpendicular to the chains, $\Delta H = 1.9$ Gauss), indicative of mobile defects [21,22]. The temperature dependence of the line-width can be described by a model of activated hopping along the chains which involves a hopping energy of ~40meV [22].

The orientation process also produces an increase of the DC conductivity [20], with stretched films showing a room temperature conductivity parallel to the chains of $3 \times 10^{-5}$ S/cm, similar to the value found for Shirakawa polyacetylene [47], and a temperature independent anistropy of about 40. Above 200K the conductivity appears activated with an energy of activation of 0.4eV. This temperature dependence will be determined by the most difficult hops; these are likely to be between chains. Thus, the temperature independent anisotropy can be explained simply by the smaller number of interchain hops in the percolation path for parallel rather than perpendicular conduction.

## 3.2. Raman spectroscopy

Resonant Raman spectroscopy has been extensively used to gain information on the strength of bond-order on polyene chains, in dehydrochlorinated PVC[23] and in poly-acetylene [24,25]. The resonantly-enhanced Raman scattering in polyacetylene shows broad phonon bands which vary in position and lineshape with the excitation energy, shifting to higher frequencies at higher excitation energies. This demonstrates that there is a distribution of bond order on the chains. There are three resonantly-enhanced Raman modes, but only those at around 1065 and 1460 $cm^{-1}$ show strong dispersion. The peaks in these two bands shift from 1065 and 1450 $cm^{-1}$ for excitation at 647.1 nm to 1124 and 1500 $cm^{-1}$ for excitation at 457.9 nm. In contrast, Durham unstretched films show a much weaker dispersion, particularly at low energies, and the limiting frequencies for excitation at 647.1 nm are 1095 and 1495 $cm^{-1}$. Stretched Durham films show behaviour more similar to that of Shirakawa material, and this is evident in figure 3 which displays the "twin-peak" structure for the two strong modes seen in "high-quality" Shirakawa material for excitation at 457.9 nm [25]; the lower frequency peak does not disperse, but the higher energy peak moves rapidly with excitation energy, merging with the lower energy feature for excitation above about 500 nm.

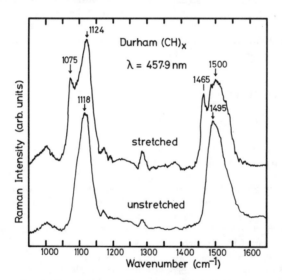

Figure 3. Raman spectra for unstretched and stretched Durham polyacetylene at 300K.

The way in which the various contributions to the bond order (Peierls distortion, chain ends etc) modify the "bare" phonon frequencies to give the spread in values measured experimentally has been most succinctly determined in the amplitude mode formalism of Horowitz and co-workers [26-30]. Analysis of the dispersion of the Raman modes in unstretched Durham polyacetylene within this model shows that the defects which limit the straight-chain lengths in the material do not impose a preferred sense of bond alternation on the chain [21, 31]. This would not be true for defects such as cis-links or $sp^3$ hybridised carbon atoms for example. Friend et al [21, 31] have suggested that the defects are chain twists or chain bends which do not break bond-alteration symmetry and hence do not prevent the formation of soliton-like states during photoexcitation or chemical doping. In contrast, if the short straight-chain sequences were terminated by defects which did impose a preferred sense of bond-alternation on the chains, polaron formation would be expected.

## 3.3.  Discussion

The increase of defect mobility along the chains in oriented Durham polyacetylene and the dispersion of the resonant-Raman spectra demonstrate that the stretching process has increased the lengths of defect-free polyene chains in the material. In contrast, unstretched films show properties consistent with chains that are frequently interrupted by defects. These defects are likely to be chain twists or bends which may be readily removed by the stretch alignment process, and are thus conformational rather than chemical in origin.

These findings demonstrate that the term "conjugation-length" needs to be applied with some care, this has not always been true in the literature. The electronic properties of the material are influenced by the extent of the straight-chain (defect-free) polyene sequences present but these are not necessarily terminated by defects which "break-conjugation". This point is very important because, as mentioned above, chains terminated by symmetry breaking defects are expected to show different properties to chains which are terminated by defects which do not break bond-alternation symmetry.


## 4.    PHOTOEXCITATION IN ORIENTED POLYACETYLENE

## 4.1.  Theory

Oriented films of Durham polyacetylene, being non-fibrous, allow us to look for anisotropic properties related to the microscopic structure of the material. In particular, it is possible to look for absorption of light with the electric vector perpendicular to the chain direction, $E_{\perp}$. This transverse absorption is not directly accessible experimentally in unoriented films where the much greater absorption coefficient, $\alpha$ for $E_{//}$ will cause almost all photons to be absorbed by intra-chain $\pi$ - $\pi^*$ transitions [32]. The spectral dependence of the absorption in unoriented films therefore shows a strong peak at the absorption edge which may be modelled by the square-root singularity in the joint optical density of states for the one-dimensional band, with some broadening due to inter-chain interactions. Measurements of $\alpha$ under pressure [33] which show a further broadening of the absorption edge provide firm evidence for the role of inter-chain tunnelling in determining the form of $\alpha$.

The close packing of chains in crystalline polyacetylene gives substantial overlap of $\pi$ orbitals on adjacent chains, and this is confirmed by three-dimensional band structure calculations, which give transverse band dispersion of the order of 0.1 eV [34]. Whilst this still gives a very considerable anisotropy between the parallel and transverse bandwidths, we emphasise that it is the absolute magnitude of the transverse transfer integral, $t_{\perp}$ which determines the rate at which electrons may move between chains. $t_{\perp}$ is of the same order as the intra-stack transfer integral in many of the metallic and superconducting charge transfer salts, and we note that for $t_{\perp} = 0.1$ eV the characteristic time for electron motion between chains, $\tau = h / t_{\perp}$ is short, $\sim 10^{-14}$ secs.

In this section, the experimental evidence for inter-chain optical excitation of electron-hole pairs with band-gap light polarised perpendicular to the chains is discussed.

## 4.2. Optical Absorption and Reflectivity.

A plot of absorption coefficient, $\alpha$, versus photon energy for an oriented polyacetylene sample is shown by the solid lines in figure 4. In this case, the sample thickness ( d ~ 3 μm) prohibited transmission measurements above 1.6 eV for $E_{//}$, however, the absorption peak has been measured on films of sub-micron thickness, stretched on suitable substrates [13], and its approximate form is shown by the broken line. It is evident from the figure that the transverse absorption is strong, a value of $\alpha_\perp = 1.0 \pm 0.2 \times 10^4$ cm$^{-1}$ being obtained at 2.2 eV [14]. As shown later, an estimate of the absorption anisotropy at

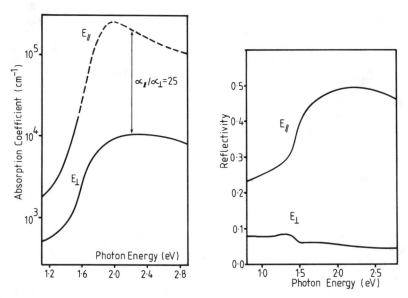

Figure 4. Optical absorption.          Figure 5. Optical reflectivity.

this energy, $\alpha_{//}/\alpha_\perp \approx 25$ can be obtained from the PA data at saturation, this gives a reasonable value for $\alpha_{//}$ of $2.5 \times 10^5$ cm$^{-1}$. Leising et al have obtained similar values for the absorption coefficients from measurements on thin films [35]. The peak value for $\alpha_{//}$ occurs at about 2 eV, which is significantly lower than the value of 2.3-2.5 eV reported for unstretched films [16], and demonstrates the effect of removal of conformational defects from the chains during orientation. Stretched films also exhibit anisotropic reflectivity as shown by figure 5, with $R_{//}$ rising to around 50% at energies above the onset of interband transitions [14].

It is possible to estimate the effects of disorder on the optical properties of the films from x-ray structure measurements which, as mentioned earlier, give a HWHM chain orientation angle to the stretch direction of 3°. Chain misorientation would thus give an apparent ratio $\alpha_{//}/\alpha_\perp = \mathrm{cosec}^2$ (3°) = 365 in the absence of transverse absorption, which is much larger than the experimental value. Moreover, there is evidence of a clear separation between $\alpha_{//}$ and $\alpha_\perp$ from thermal modulation spectra ($\partial\alpha/\partial\theta$ vs photon energy) [14] and Raman excitation profiles for $E_{//}$ and $E_\perp$ [36], which demonstrate that the parallel and transverse optical absorption processes involve different electronic transitions.

## 4.3    Photoinduced Absorption and Photoconductivity

The excitation source used for PA experiments discussed here was a Nd:YAG and dye laser combination, producing 12 nsec pulses at 2.2eV. The change in sample transmission, $\Delta T$ was obtained either by integrating the detector signal with a lock-in amplifier or, for time-resolved measurements, by using a fast digital oscilloscope [14].

The PA spectrum of oriented Durham polyacetylene obtained at 20K (illustrated in figure 6) shows a single asymmetric absorption feature for parallel polarisation of the probe beam with interference structure near the peak at 0.48 eV. This feature is considered to be characteristic of charged soliton, $S^{\pm}$ defects [37,38]. PA due to these states is observed with both polarisations of the excitation beam, but is more strongly induced by $E_{\perp}$ excitation, the anisotropy ratio being intensity dependent, as shown in figure 7. In contrast to Shirakawa polyacetylene [39], the PA feature shows an initially linear dependence with laser intensity, with saturation at high intensities. At saturation, presuming that the density of photogenerated defects within the absorption depth, $1/\alpha$ is the same for $E_{//}$ and $E_{\perp}$, we simply probe the ratio $\alpha_{//}/\alpha_{\perp}$, which we find to be $\approx 25$ at 2.2eV. In contrast, at intensities where the response is linear for both polarisations, the true anisotropy in the efficiency of photogeneration of defect states is measured, and we find a value of $\approx 4$ in favour of excitation with $E_{\perp}$.

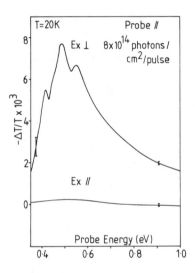

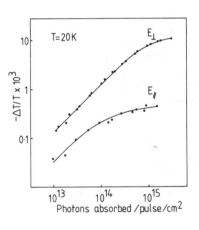

Figure 6.  PA spectra with probe parallel to chains.

Figure 7.  Intensity dependence of the PA near 0.5 eV.

Figure 8 illustrates the time dependence of the PA at 0.77 eV at three different temperatures, the inset shows the temperature dependence of the integrated signal. The defect states are long-lived, showing a decay time of about 100 msec. The decrease in magnitude of the PA as the temperature increases has been attributed to a decrease in the lifetime of the photoinduced states [40,41], however, figure 8 shows a temperature independent decay on the time scale of $\approx 1$ to 100 msecs, hence, the temperature dependence is determined mainly by the initial fast decay. Further time resolved measurements carried out at 1.65 eV reveal a similarly long-lived signal caused by bleaching of the interband transition which can be associated with the defects states. Thus in Durham polyacetylene the $S^{\pm}$ defects obtain their oscillator strength from the band states [14] and not from $S^0$ levels near the band edge as has been suggested for Shirakawa polyacetylene [42].

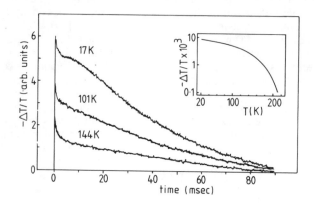

Figure 8. Time dependence of PA at 0.77 eV. Inset shows temperature dependence of integrated signal.

We have found that the photoconductivity of stretched films is only measureable below about 180K, the apparent photocurrent at higher temperature being due to thermal modulation of the dark current [20]. At low temperatures the photocurrent is dominated by a short-lived component, as is found for Shirakawa material [43], our measurements indicate that this component is relatively isotropic (similar results have recently been obtained by Bleier et al [44]). The intensity dependence of the PC also indicates that a fraction of the total photocurrent measured is due to the $S^{\pm}$ states observed in PA experiments. We find that at 170K, for laser intensities below the saturation regime, this fraction is around 50% [14]. Work is presently underway to obtain more information on the PC.

## 4.4.  Discussion

The experimental results discussed in 4.2 and 4.3 imply that a substantial fraction of $\alpha_{\perp}$ is due to inter-chain transitions which create electron-hole pairs initially on separate chains. Theoretical studies [45] are in agreement with this model, and predict that $\alpha_{\perp}$ is dominated by inter-chain transitions near the bandgap and by intra-chain transitions further into the band. We consider that carriers separated between chains are more likely to survive as long-lived excitations than intra-chain electron-hole excitations. The latter, through confinement to a one-dimensional chain (of finite length) and the Coulomb interaction, not considered in the SSH hamiltonian [46], must be very likely to decay by geminate recombination. This mechanism therefore provides a natural explanation for the preferential excitation of $S^{\pm}$ defects by light polarised perpendicular to the chains. Inter-chain excitation of electron-hole pairs must, in the first instance, produce positive and negative polarons, $P^{\pm}$, however, in the time scale probed in our experiments these must have combined to form pairs of similarly charged solitons on a chain, since the spectral feature measured in the PA experiments is characteristic of soliton defects with a single level in the gap.

## 5. CONCLUSIONS

The synthesis of highly-oriented films of polyacetylene via the Durham precursor route has allowed an investigation of the intrinsic anisotropic properties of the material. In particular, we have been able to study the role of inter-chain transitions in determining the photoexcitation properties of the material and find that long-lived photocarriers are preferentially excited by such transitions.

In this review I have covered some of the recent work on the electronic properties of Durham polyacetylene carried out at Cambridge. Further work is currently underway on this and other conjugated polymers.

ACKNOWLEDGEMENTS

I would like to thank R.H.Friend and D.D.C.Bradley for many helpful discussions and St. John's College, Cambridge for a research fellowship. Work on Durham Polyacetylene at Cambridge is supported by British Petroleum p.l.c.

## References

1.  H Munstedt - *Solid State Sciences* (Springer, New York), **63**, 8 (1985).

2.  W J Feast - *Handbook of Conducting Polymers*, ed. T Skotheim (M Dekker, New York) **1**, 1 (1986).

3.  J H Edwards & W J Feast - *Polymer Commun.*, **21**, 595 (1980).

4.  D C Bott, C K Chai, J H Edwards, W J Feast, R H Friend & M E Horton - *J. Phys. (Paris)*, **44**, C3, 143 (1983).

5.  R H Friend, D C Bott, D D C Bradley, C K Chai, W J Feast, P J S Foot, J R M Giles, M E Horton, C M Pereira & P D Townsend - *Phil. Trans. Roy. Soc.*, **A314**, 37 (1985).

6.  D C Bott, C S Brown, C K Chai, N S Walker, W J Feast, P J S Foot, P D Calvert, N C Billingham & R H Friend - *Synthetic Metals*, **14**, 245 (1986).

7.  K Harper & P G James - *Mol. Cryst. Liq. Cryst.*, **117**, 95 (1985).

8.  M E Horton, D D C Bradley, R H Friend, C K Chai & D C Bott - ibid, 51 (1985).

9.  G Leising - *Polym. Bull.*, **11**, 401 (1984).

10. M M Sokolowski, E A Marseglia & R H Friend - *Polymer* (in press).

11. H Kahlert & G Leising - *Mol. Cryst. Liq. Cryst.*, **117**, 1 (1985).

12. H Kahlert, O Leitner & G Leising - ICSM Kyoto (1986), *Synthetic Metals* (in press).

13.  G Leising, H Kahlert & O Leitner - *Solid State Sciences* (Springer, New York), **63**, 56 (1985).

14.  P D Townsend & R H Friend - ICSM Kyoto (1986), *Synthetic Metals* (in press).

15.  P J S Foot, P D Calvert, M Ware, N C Billingham & D C Bott - *Mol. Cryst. Liq. Cryst.*, **117**, 47 (1985).

16.  R H Friend, D D C Bradley, C M Pereira, P D Townsend, D C Bott & K P J Williams - *Synthetic Metals*, **13**, 101 (1986).

17.  D D C Bradley, R H Friend, T Hartmann, E A Marseglia, M M Sokolowski & P D Townsend - ICSM Kyoto (1986), *Synthetic Metals* (in press).

18.  H Kahlert, O Leitner & G Leising - ibid.

19.  P D Townsend, C M Pereira, D D C Bradley, M E Horton & R H Friend - *J. Phys. C*, **18**, L283 (1985).

20.  P D Townsend, D D C Bradley, M E Horton, C M Pereira, R H Friend, N C Billingham, P D Calvert, P J S Foot, D C Bott, C K Chai, N S Walker & K P J Williams - *Solid State Sciences* (Springer, New York), **63**, 50 (1985).

21.  R H Friend, D D C Bradley, C M Pereira, P D Townsend, D C Bott & K P J Williams - *Synthetic Metals*, **13**, 101 (1986).

22.  M E Horton, R H Friend, P J S Foot, N C Billingham & P D Calvert - ICSM Kyoto (1986), *Synthetic Metals* (in press).

23.  A Baruya, D L Gerrard & W F Maddams - *Macromolecules*, **16**, 578 (1983).

24.  H Kusmany - *Phys. Stat. Solidi (b)*, **97**, 521 (1980).

25.  D B Fitchen - *Mol. Cryst. Liq. Cryst.*, **83**, 95 (1982).

26.  B Horovitz - *Solid State Commun.*, **41**, 729 (1982).

27.  Z Vardeny, E Ehrenfreund, O Brafman & B Horovitz - *Phys. Rev. Lett.*, **51**, 2326 (1983).

28.  B Horovitz, Z Vardeny, E Ehrenfreund & O Brafman - *Synthetic Metals*, **9**, 215 (1984).

29.  Z Vardeny, E Ehrenfreund, O Brafman & B Horovitz - *Phys. Rev. Lett.*, **54**, 75 (1983).

30.  E Ehrenfreund, Z Vardeny, O Brafman & B Horovitz - (preprint).

31.  R H Friend, D D C Bradley, P D Townsend & D C Bott - ICSM Kyoto (1986), *Synthetic Metals* (in press).

32.  C R Fincher, M Osaki, M Tanaka, D Peebles, L Lauchlan & A J Heeger -
     *Phys. Rev. B* **20**, 1589 (1979).

33.  D Moses, A Feldblum, E Ehrenfreund, A J Heeger, T C Chung & A G MacDiarmid
     *Phys. Rev. B* **26**, 3361 (1982).

34.  P M Grant & I P Batra - *J. Phys. (Paris)*, **44**, C3, 437 (1983).

35.  G Leising & H Kahlert - (private communication).

36.  P Knoll, G Leising & H Kusmany - (to be published).

37.  G B Blanchet, C R Fincher, T C Chung & A J Heeger - *Phys. Rev. Lett.*, **50**,
     1938 (1983).

38.  J Orenstein, Z Vardeny, G L Baker, G Eagle & S Etemad - *Phys. Rev. B* **30**, 786
     (1984).

39.  Z Vardeny, J Orenstein & G L Baker - *J. Phys. (Paris)*, **44**, C3, 325 (1983).

40.  Z Vardeny, J Orenstein & G L Baker - *Phys. Rev. Lett.*, **50**, 2032 (1983).

41.  Z Vardeny, J Strait, D Moses, T C Chung & A J Heeger - *Phys. Rev. Lett.*, **49**,
     1657 (1982).

42.  J Orenstein - *Handbook of Conducting Polymers*, ed. T Skotheim (M Dekker,
     New York), **2**, 1297 (1986).

43.  M Sinclair, D Moses & A J Heeger - (preprint).

44.  H Bleier, G Leising & S Roth - ICSM Kyoto (1986), *Synthetic Metals* (in
     press).

45.  P L Danielsen - *J. Phys. C Lett.* (submitted for publication).

46.  W P Su & J R Schrieffer - *Proc. Natl. Acad. Sci. USA*, **77**, 5626 (1980).

47.  Y W Park, A J Heeger, M A Druy & A G MacDiarmid - *J. Chem. Phys.*, **15**, 946
     (1980).

IONICALLY-CONDUCTIVE SOLID SOLUTIONS OF DIVALENT CATION SALTS IN
POLY(ETHYLENE OXIDE)

R. Huq, L-L. Yang and G. C. Farrington
Department of Materials Science
University of Pennsylvania
3231 Walnut Street
Philadelphia, PA 19104

ABSTRACT.  Poly(ethylene oxide) (PEO) forms solid solutions with $MgCl_2$,
$PbI_2$, $PbBr_2$, and various other salts of divalent cations.  DSC analysis and
a study of the temperature dependence of conductivity indicate that these
materials consist of several crystalline phases, corresponding to pure PEO
and salt-rich complexes, and a coexisting elastomeric phase.  $MgCl_2 \cdot (PEO)_{16}$
has an ionic conductivity comparable to that of $LiCF_3SO_3 \cdot (PEO)_{96}$ from $10^{-5}$
$(ohm \cdot cm)^{-1}$ at 80°C, and the conductivity of $PbBr_2 \cdot (PEO)_8$ is $10^{-6}$-$10^{-7}$ (ohm
$cm)^{-1}$ at 180°C and about $10^{-5}$ $(ohm \cdot cm)^{-1}$ at 250°C.  The lead halide complexes conduct both $Pb^{2+}$ and halide anions.  All of the divalent halide
complexes are stable to nearly 300°C.

1.  INTRODUCTION

Wright et al. [1] and Armand [2] first showed that poly(ethylene
oxide)(PEO) and various alkali metal salts form solid solutions which have
significant ionic conductivities.  Their work stimulated interest in these
non-crystalline solid electrolytes and their potential use as separators in
solid-state batteries [1-6].  Several large programs aimed at developing
batteries of this type using Li anodes are underway.

Unlike most crystalline solid electrolytes, the PEO-based electrolytes, such
as $LiClO_4 \cdot (PEO)_8$, conduct both anions and cations.  While many high-
conductivity solid electrolytes for cations are known, there are relatively
few good conductors of anions.  This work began as an exploration of the
possibility of preparing PEO electrolytes in which the anion transport
number is essentially one.  The initial strategy was to prepare solid solutions of divalent cations and monovalent anions in PEO, with the
expectations that the divalent cations would be trapped and their diffusion
impeded by their double charge.

Solutions of divalent cation salts in polymers such as PEO have not been
widely investigated.  James et al. [7] prepared complexes of divalent cations such as the halides of Zn, Co, Fe, and Cu, with poly(propylene oxide)
which formed single-phase amorphous compositions.  They studied the thermal
behavior and mechanical properties of these materials, but did not report
any conductivity measurements.  Fontanella et al. [8] recently prepared

89

L. Alcácer (ed.), Conducting Polymers, 89–93.

$Ca(SCN)_2 \cdot 4H_2O \cdot (PEO)_{6.5}$ and $Ba(SCN)_2 \cdot 3H_2O \cdot (PEO)_{6.5}$ and found them to form films with high glass transition temperatures and ionic conductivities lower than that of pure PEO. Since beginning our work, we have learned that Patrick et al. [9] and Abrantes et al. [10] have also been investigating the preparation and properties of electrolytes formed by PEO and divalent cation salts.

We have found that PEO easily forms solid solutions with various divalent halides, including $MgCl_2$, $PbBr_2$, and $PbI_2$. Differential scanning calorimetry (DSC), thermogravimetric analysis (TGA), x-ray and electron microscopy, and complex AC impedance analysis were used to study the composition, stability, and conductivity of these materials. Some have high conductivities at moderate temperatures. For example, the ionic conductivity of $MgCl_2 \cdot (PEO)_{16}$ is about $10^{-5}$ $(ohm\text{-}cm)^{-1}$ at 80°C, comparable to that reported for $LiCF_3SO_3 \cdot (PEO)_9$. As initially expected, the solid solutions of PEO with salts of Mg(II) appear to be principally conductors of anions. However, simple transport number measurements indicate that the solutions of Pb(II) salts also conduct Pb(II).

## 2. ELECTROLYTE PREPARATION AND CHARACTERIZATION

### 2.1 $MgCl_2$-PEO Electrolytes

A family of $MgCl_2 \cdot (PEO)_n$ compositions in which n=4,8,12,16, and 24 and PEO is a symbol for $(CH_2\text{-}CH_2O)$ have been prepared by solution casting. Appropriate quantities of dried PEO (Polysciences, MW $5 \times 10^6$) and $MgCl_2$ were dissolved in acetonitrile and anhydrous ethanol, respectively. The solutions of PEO and $MgCl_2$ were then mixed to form a homogeneous solution with the desired molar ratio of O/Mg and a total concentration of about 4% by weight. This solution was cast within glass rings on polished Teflon plates to produce final film thicknesses of 50-100 microns. The solvent were first slowly evaporated at room temperature in a desiccator over molecular sieves, then evacuated to $10^{-2}$ torr and heated gradually from room temperature to 100°C. Finally, films were held at 100°C for 24 hours. The resulting films were quite hygroscopic and were stored in a drybox to minimize water absorption.

Thermogravimetric analysis of a film of $MgCl_2 \cdot (PEO)_{16}$ that was exposed to the air revealed a large weight loss between 25 and 70°C. From 100 to 250°C, no additional weight loss was observed, and the sample finally decomposed above 290°C. Heating and cooling cycles demonstrated that the low temperature weight loss is the result of water desorption. DSC and x-ray diffraction studies found no evidence of free $MgCl_2$ in the dry films.

## 2.2  Conductivity Studies

AC conductivity measurements from $10^2$ to $10^6$ Hz were made with a Solartron
1170 Frequency Response Analyzer and polished brass electrodes.  The room
temperature conductivities of $MgCl_2.(PEO)_n$ electrolytes are rather low,
below $10^{-8}$ $(ohm-cm)^{-1}$.  As the samples are heated, the more dilute
materials, n =12, 16, and 24, showed a clear transition to higher conductivities around 60°C, the temperature at which pure PEO melts.  Similar
behavior has been observed with complexes of alkali metal salts.  This
transition is suppressed at higher $MgCl_2$ concentrations (n=4,8).  Ionic
conductivities of $10^{-5}$ $(ohm-cm)^{-1}$ are reached by 80°C.

The conductivities of films of composition n=12, 16, and 24 are all quite
similar.  The composition with the highest conductivity above the PEO melting point is n=16.

## 2.2  Stability Studies

DSC analyses from room temperature to 300°C on samples of $MgCl_2.(PEO)_n$ in
which n=4, 8, 16, and 24 indicate that two crystalline phases coexist in the
$MgCl_2$.PEO complexes at room temperature: pure PEO and the salt-rich complex.
These two phases produce two endotherms.  The first peak is sharp and occurs
about 60°C.  It can be attributed to the melting of the pure PEO crystalline
phase.  The second is more diffuse and starts above the PEO melting peak and
extends to about 200°C.  We have attributed it to the melting and dissolution of the salt-rich crystalline complex.  The fractions of these
crystalline phases vary with O/Mg (n) ratio in the film.  The largest amount
of PEO and smallest amount of the salt complex exist, as would be expected,
in the most dilute complex, $MgCl_2.(PEO)_{24}$.

The crystalline phases are of little importance for ionic conductivity,
which appears to occur in the elastometric solid solution of salt in PEO.
The thermal studies presented here suggest that both the fraction of the
elastomeric conducting phase and its salt concentration for a given composition depend on the overall concentration of salt and on the thermal history
of the electrolyte.  At high salt concentrations (n=4,8), salt-PEO complex
formation is favored and the concentration of free PEO is small.
Consequently, the major glass transition temperatures are high, the films
are not flexible at lower temperatures, and their ionic conductivities are
quite low.  At lower salt concentrations, the concentration of free PEO is
high, but the concentration of ions is smaller, and the conductivity
decreases.  The conductivity is largest at intermediate compositions, such
as $MgCl_2.(PEO)_{16}$.

## 2.3  $PbI_2$ and $PbBr_2$-PEO Electrolytes

Lead halide-PEO electrolytes were also prepared from mixed solvents.  PEO
was dissolved in acetonitrile, and the lead halides in dimethylsulfoxide
(DMSO).  As before, films were cast on Teflon plates and then placed in a
vacuum oven.  They were first vacuum dried at room temperature for 24 hours,
and then slowly heated.  They were maintained under vacuum at 55°C for two
days, after which they easily peeled from the Teflon plate.

The $PbI_2$ complexes were yellow or yellow-orange, depending on composition, and the lead bromides were colorless. In general, the films were flexible and semi-transparent.

## 2.4 Conductivity Measurements

Polymer film conductivities were measured by complex AC impedance analysis, as described previously for $MgCl_2$ electrolytes. Among the lead bromide electrolytes, $PbBr_2.(PEO)_8$ has the highest conductivity between 60°C and 200°C. It is about $10^{-6}$ $(ohm-cm)^{-1}$ at 180°C. The variation of conductivity with temperature cycling reveals complex hystereses that surely are the result of the various dissolution and crystallization processes which occur as the temperature is changed. The lead iodide electrolytes are generally lower in conductivity compared to the bromides.

## 2.4 Stability Studies

DSC traces on the lead halide electrolytes showed no sample degradation below 300°C in argon. An endotherm resulting from the melting of PEO was observed, but there was no evidence of well-defined melting points of any complexes of the lead salts. Between the melting of PEO at 65°C and the final decomposition of the samples, a broad endotherm was observed which can be associated with an extended process of melting and dissolution of lead halide complex compositions in PEO.

TGA studies on freshly-prepared films showed a gradual weight loss from room temperature to about 200°C which was not seen on subsequent heatings. It can be associated with the loss of water and residual solvent. No further weight loss was observed in dry nitrogen until the electrolytes irreversibly decomposed around 310°C.

## 2.5 Transport Number Measurements

DC and AC conductivity analysis on the Mg(II) and Pb(II) electrolytes were carried out using non-blocking (Mg or Pb) and blocking electrodes. The Mg(II) electrolytes showed no evidence of Mg(II) motion and appear to be virtually pure anion conductors. The Pb(II) electrolytes appeared to be good conductors of Pb(II) as well as halide ions. An initial estimate of the transport number of Pb(II) in $PbBr_2.(PEO)_{20}$ is 0.6-0.7 at 140°C. We must caution that these transport number measurements are preliminary estimates. It is a major undertaking to measure definitive transport numbers, and that work is not yet begun.

## 3. CONCLUSIONS

It is clear from this work that PEO easily forms solid solutions with a variety of salts of Pb(II) and Mg(II) over a wide range of compositions. Continuing investigations have also shown that similar PEO-based polymer electrolytes can be formed with halides of Ca(II), Ba(II)), Cd(II), Ni(II), Zn(II), and other divalent cations [11].

Some of these materials conduct both anions and cations, others are principally conductors of anions. It appears that solutions of small, polarizing ions, such as Mg(II) and Ca(II), are anion conductors. The cations are presumably trapped in strong associations with oxygen ions in the PEO chains, while the larger anions are less-strongly solvated by the polymer. In contrast, divalent cations that are larger and less polarizing (more 'covalent'), such as Pb(II), may themselves be as mobile as the anions. These differences are logical and not unexpected in light of conventional concepts of ion/solvent interactions.

Thermal analysis has shown that PEO electrolytes formed with divalent cation salts are complex mixtures of crystalline and amorphous phases. At most temperatures they are heterogeneous electrolytes in which high conductivity is associated with the amorphous, elastomeric regions and low conductivity with the crystalline phases.

When the PEO electrolytes were first investigated, they were frequently considered relatives of crystalline solid electrolytes. In fact, they more-closely resemble classical solutions of salts in non-aqueous electrolytes with very high viscosities. They are semi-solid solutions whose properties and applications lie between those of true liquids and crystalline solids. As such, they represent an intriguing and possibly useful area of non-aqueous electrochemistry that has received relatively little systematic investigation.

## 4. ACKNOWLEDGEMENTS

This work was supported by the Assistant Secretary for Conservation and Renewable Energy, Office of Energy Systems Research, Energy Storage Division of the U.S. Department of Energy under Contract No. DE-AC02-76SF00098. Additional support from the National Science Foundation, MRL Program under Grant DMR 821-6718 is gratefully appreciated.

## 5. REFERENCES

1. B. E. Fenton, J. M. Parker and P. V. Wright, Polymer, 14 (1973) 589.
2. M. Armand, J. M.. Chabagno and M. Duclot, in Fast Ion Transport in Solids, P. Vashishta, J. N. Mundy and G. K. Shenoy, eds., (North-Holland, New York, 1979) p. 131.
3. W. I. Archer and R. D. Armstrong, ELectrochim. Acta, 25 (1980) 1689.
4. D. F. Shriver, B. L. Papke,, M. A. Ratner, R. Dupon, T. Wong and M. Brodwin, Solid State Ionics, 5 (1981) 83.
5. J. E. Weston and B.C.H. Steele, Solid State Ionics, 2 (1981) 347.
6. J. E. Weston and B.C.H. Steele, Solid State Ionics, 7 (1982) 81.
7. D. B. James, R. E. Wetton and D. S. Brown, Polymer, 20 (1979) 187.
8. J. J. Fontanella, M. C. Wintersgill and J. P. Calame, J. Polymer Sci., Polym. Phys. Ed., 23 (1985) 113.
9. A. Patrick, M. Glasse, R. Latham and R. Linford, Solid State Ionics, accepted for publication.
10. T.M.A. Abrantes, L. J. Alcacer and C.A.C. Sequeira, Solid State Ionics, accepted for publication.
11. L-L. Yang, H. Yang, R. Huq and G. C. Farrington, University of Pennsylvania, unpublished results.

ROOM TEMPERATURE HIGH IONIC CONDUCTIVITY FROM ALKALI METAL - SILVER
HALIDE - POLY(ETHYLENE-OXIDE) COMPLEXES

J.R. Stevens  
Physics Department  
University of Guelph  
Guelph, Ontario  
Canada  N1G 2W1

B.E. Mellander  
Physics Department  
Chalmers University of  
Technology  
S-41296 Goteborg, Sweden

ABSTRACT

Rb $Ag_4I_5$ and $KAg_4I_5$ in a freshly prepared state have unusually high
ionic conductivity at room temperature. Unfortunately these salts are
unstable against disproportionation. Stable complexes of the form
poly(ethylene oxide) - M $Ag_4I_5$ (M = Li, K, Rb) have been prepared and
their ionic conductivities measured for various M:0 ratios. For
example for K:0 = 1:1 the conductivity is $2 \times 10^{-3}$ $ohm^{-1}$ $cm^{-1}$ at room
temperature and the activation energy is 0.16 eV. Studies using
differential scanning calorimetry, NMR and wide angle x-rays have been
made and show that these complexes are polycrystalline but with no
evidence of crystalline poly(ethylene-oxide). We suggest that the long
polymeric chains inhibit the attainment of the positive activation
volume required for disproportionation.

1.  Introduction

The two solid electrolytes $RbAg_4I_5$ and $KAg_4I_5$ have unusually high ionic
conductivity near room temperature, about 0.2 $\Omega^{-1}$ $cm^{-1}$ for both
compounds in a freshly prepared state [1-8]. These conductivities
increase with temperature with activation energies of 0.12 eV [1,4,5].
Unfortunately, these compounds are unstable to disproportionation,
$RbAg_4I_5$ at temperatures below 27°C [9] and $KAg_4I_5$ at temperatures below
38°C [1]. In spite of this thermodynamic instability it is reported
that $RbAg_4I_5$ may be retained at room temperature for indefinite periods
of time in a dry atmosphere [5,10-12]. $KAg_4I_5$ is less stable but only
qualitative estimates of this stability have been reported [5,11].
Both of these salts are hygroscopic and water catalyses the
disproportionation. The objective of the work reported here was to
incorporate AgI into complexes which would prove to be stable while
maintaining high ionic conductivity. Because of all the attention
being paid to poly(ethylene oxide) (PEO) [13,14] it seemed appropriate
to incorporate the polymer in some way.

95

L. Alcácer (ed.), Conducting Polymers, 95–102.
© 1987 by D. Reidel Publishing Company.

2.  Preparation

AgI is insoluble in organic solvents.  Early preparation of $RbAg_4I_5$ and $KAg_4I_5$ was by solid state reaction; for example stoichiometric amounts of AgI and RbI were combined and melted.  AgI, RbI and KI are soluble in HI at about $50^{\circ}C$ and crystals may be grown from solution [11].  In this work $MAg_4I_5$ (M = Li, K, Rb) were chemically complexed with high molecular weight PEO in a nitrogen atmosphere with concentrations of M:O of 0.1, 0.2, 0.4, 0.6 and 1.0.  Stoichiometric amounts (exact to the milligram) of PEO, AgI and MI of 99.9% purity were mechanically mixed and then combined with water-free acetonitrile in the ratio of 1:25 by volume.  The PEO was obtained from Polysciences Inc. and had a weight-average molecular weight of 600000.  The mixture was heated to $50^{\circ}C$ and stirred.  Within a short time the compound $P(EO)_{0:M}MAg_4I_5$ precipitated out and the solvent was then evaporated off.  Depending on the errors in the stoichiometry traces of AgI and MI could be found. Precision stoichiometry is important because of the nature of the phase diagram [1,2] for the salt $M_xAg_yI_z$ (M=Rb, K).  The solid electrolytes are formed precisely on the 20 mole% MI, 80 mole% AgI line which exists between $27^{\circ}C$ and $228^{\circ}C$ for $RbAg_4I_5$ and between $38^{\circ}C$ and $253^{\circ}C$ for $KAg_4I_5$.  Because of the insolubility of AgI in acetonitrile we were not able to use this method to form AgI-PEO complexes.  We could not find a reliable phase diagram for the LiI-AgI system in the literature.

3.  Experimental

The samples were pressed in a cylindrical die of 13 mm diameter using a pressure of 0.6 GPa.  The samples were 1 to 2 mm long and perforated platinum or silver foils pressed on the sample surfaces served as electrodes.  These electrodes were satisfactory in the short term but eventually pealed off.  Electrodes of silver paint were found to be most stable in the long term.  The samples were placed between two gold plates under spring pressure.  The sample holder was placed inside a furnace and the sample was kept in a dry nitrogen atmosphere.  The electrical conductivity was determined using complex impedance analysis.  The complex impedance was measured using a computer controlled HP2474A LCR meter.  The frequency range covered was 100 Hz and the applied signal was 20 mV.  A typical impedance plot is shown in fig. 1.  The sample resistance was obtained from the intercept of the straight line and the real axis.  The DCS measurements were performed using a Rigaku D DSC apparatus.  The samples were contained in platinum cups; a protective atmosphere of dry nitrogen gas was used.

The results of room temperature measurements of conductivity are presented in fig. 2 as $\log_{10}\sigma$ versus the M:O ratio (M=K, Rb).  It should be noted here that each K or Rb ion is accompanied by four Ag ions.  Thus, if the ratios above are presented as metal ion to oxygen ratios they will be five times as high as those in fig. 2.  For K:O ratios less than 0.2 the conductivity decreases rapidly and the trend

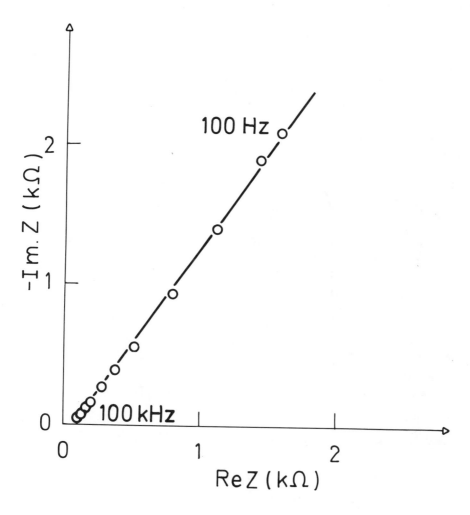

Fig. 1   Complex impedance plot for a $P(EO)_1KAg_4I_5$ sample with silver
         electrodes.


is   similar   for   the   samples   containing   rubidium.      The   best
conductivities for samples containing either potassium or rubidium salt
are thus about $2 \times 10^{-3}\ \Omega^{-1}\ cm^{-1}$.   For $P(EO)_1LiAg_4I_5$, $\sigma = 1 \times 10^{-4}\ \Omega^{-1}$
$cm^{-1}$ at 20°C.   In fig. 3 the temperature dependence of the conductivity

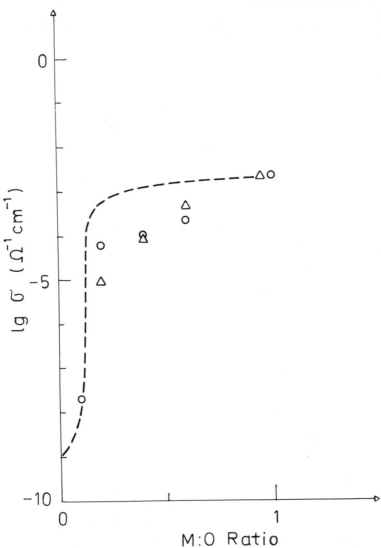

Fig. 2   Room temperature electrical conductivity of $P(EO)_{O:M}MAg_4I_5$
         for M = K (circles) and M = Rb (triangles).   The dashed curve
         shows the calculated conductivity using percolation theory.

is shown as a $\ln(\sigma T)$ versus $1/T$ plot for the $P(EO)_1KAg_4I_5$ sample.   The
activation energy is, in this case, 0.16 eV.   The electronic
contribution to the electrical conducitivity has been measured using a
dc-polarization method.   For all samples tested the electronic
conductivity is negligible compared to the ionic conductivity.

The DSC measurements showed that a number of different phases could be detected in the samples. In fig. 4 some examples of DSC plots are shown. For pure PEO a large peak at about 60°C indicate the melting point of the polymer. This peak was not found in any of the samples where silver salts were present. Furthermore, in all samples, the peak corresponding to the phase transition at 147°C in pure AgI was detected. In addition to the peak at 253°C for $KAg_4I_5$ and the peak at 228°C for $RbAg_4I_5$ peaks at 238°C and 197°C indicated the nonstoichiometry of the samples, compare the phase diagram [1,2]. One broad peak that could not be explained as a silver salt peak was observed at about 160°C.

High reslution, solid state NMR measurements on pure PEO and $P(EO)_1KAg_{45}$ confirm the results of the DSC measurements that the crystalline form of PEO does not remain in the samples where the PEO has been mixed with the salt. $T_1$ measurements using $^{109}Ag$ indicate that the Ag ions are very mobile. More will be reported on these measurements later. These results are consistent with those of Looser et al [5] on the pure salts $MAg_4I_5$.

Wide angle X-ray diffraction measurements were made on the salt - PEO complexes. The spectrum is shown in fig. 5 for $P(EO)_1$ $KAg_4I_5$. In addition to the $KAg_4I_5$ peaks which predominate there are peaks due to AgI (confirming the DSC results) and due to $K_2AgI_3$ and Pt. The $K_2AgI_3$ is present because of the inaccurate stoichiometry and not because of disproportionation. At temperatures between 38°C and 130°C a slight increase in the mole% KI above 20% will provide a small amount of $K_2AgI_3$ in addition to $KAg_4I_5$. $K_2AgI_3$ is yellow and non conducting. $KAg_4I_5$ is white and conducting. There is no evidence of crystalline PEO in the X-ray diffraction spectrum.

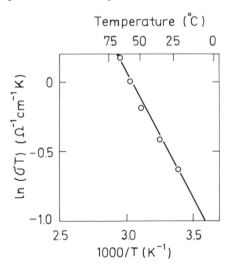

Fig. 3    $ln(\sigma T)$ versus $1000/T$ plot for a $P(EO)_1KAg_4I_5$ sample.

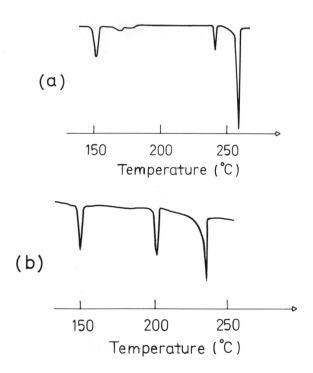

Fig. 4    DSC plots for. (a) P(EO)$_1$KAg$_4$I$_5$ and (b) P(EO)$_1$RbAg$_4$I$_5$.

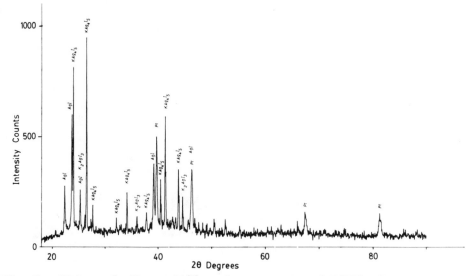

Fig. 5    Wide angle X-ray diffraction spectrum of P(EO)$_1$ KAg$_4$I$_5$.

## 4. Discussion

The samples thus contain crystalline salt phases and including PEO-salt complex. We conclude that the high conductivity is due to the presence of $MAg_4I_5$ since $\sigma$ decreases with decreasing $MAg_4I_5$ content and since the activation energy is very close to that for the pure $MAg_4I_5$ salt. Usually PEO-salt complexes have higher activation energies [14]. It has been shown that the electrical conducitivity of some two-phase systems can be described using effective medium percolation theory [14,15]. Even if the present case is more complicated it can be of interest to compare the position of the percolation threshold with the rapid drop in conductivity observed for M:O ratios of about 0.1, see fig. 2. The theoretical conductivity of a two-phase system, salt and polymer can be calculated using percolation theory. Normalizing on the highest conducitivity for the complex $(3 \times 10^{-3}(\Omega \ cm)^{-1})$ and using a value of $1 \times 10^{-9}$ $(\Omega \ cm)^{-1}$ for the polymer the theoretical conductivity as a function of M:O ratios follows the dashed line in fig. 2. It is interesting to note that the position of the percolation threshold, which is not sensitive to the conductivity values chosen, agrees well with the rapid drop in conductivity found for M:O ratios of about 0.1. The complexes being discussed here are more complicated than a simple two-phase mixture since traces of both AgI and MI have been detected and the behaviour of the polymer is not known. The position of the percolation threshold indicates, however, that percolation phenomena has to be taken into account in this case.

The conductivity of $P(EO)_1 KAg_4I_5$ has not changed for more than a year. We suggest that the attainment of the positive activation volume required for disproportionation is prevented by the long chain PEO molecules. $LiAg_4I_5$ quickly discolors as it picks up water from the air and disproportionates. However, $P(EO)_1LiAg_4I_5$ left in the open air discolored only slightly over one month.

We have not been successful preparing samples with high $\sigma$ at room temperature using methanol as a solvent. The reason for this is not clear to us. We have examined $P(EO)_{0.M}KAg_4I_5$ compounds prepared using methanol and acetonitrile for C, H and N by elemental analysis. Within the accuracy of the measurements there was not difference between compounds prepared using different solvents. The results reported here are preliminary and work will continue, especially in regard to long term thermodynamic stability.

## Acknowledgement

We would like to thank Professor A. Lundén for continuous support and valuable discussions. We would also like to thank T. Hjertberg of the Polymer Research group for performing the solid state, high resolution NMR measurements and K. Jeffrey of the University of Guelph for doing the $T_1$ - NMR measurements. This work has been supported financially by the Swedish Natural Sciences Reseach Council and the Swedish Board of Technical Development which is gratefully acknowledged.

References

[1]   J.B. Bradley and P.D. Greene, Trans. Faraday Soc. 62 (1966)
      2069.
[2]   J.N. Bradley and P.D. Greene, Trans. Faraday Soc. 63 (1967)
      424.
[3]   B.B. Owens and G.R. Argue, Science 157 (1967) 308.
[4]   B.B. Owens and G.R. Argue, J. Electrochem. Soc. 117 (1970) 898.
[5]   H. Looser, M. Mali, J. Roos and D. Brinkmann, Solid State Ionics
      9/10 (1983) 1237.
[6]   D.O. Raleigh, J. Appl. Phys. 41 (1970) 1876.
[7]   B.B. Owens, in: Advances in electrochemistry and electrochemical
      engineering, ed. C.W. Tobias, Vol. 8 (Wiley, New York, 1971).
[8]   B. Scrosati, J. Appl. Chem. Biotechnol. 21 (1971) 223.
[9]   L.E. Topol and B.B. Owens, J. Phys. Chem. 72 (1968) 2106.
[10]  W.V. Johnston, H. Wiedersich and G.W. Lindberg, J. Chem. Phys.
      51 (1969) 3739.
[11]  D.A. Gallagher and M.V. Klein, Phys. Rev. B 19 (1979) 4282.
[12]  B.B. Owens, B.K. Patel, P.M. Skarstad and D.L. Warburton, Solid
      State Ionics 9/10 (1983) 1241.
[13]  M. Armand, Solid State Ionics 9/10 (1983) 745.
[14]  G.T. Davis and C.K. Chiang, in: Conducting polymeric materials,
      ed. H. Sasabe (Chemical Marketing Co., Tokyo, 1984) p. 244.
[15]  D.G. Ast, Phys. Rev. Letters 33 (1974) 1042.
[16]  M.F. Bell, M. Sayer, D.S. Smith and P.S. Nicholson, Solid State
      Ionics 9/10 (1983) 731.

# POLYMERS WITH BOTH IONIC AND ELECTRONIC CONDUCTIVITY

J. Owen
Department of Chemistry and Applied Chemistry
University of Salford
UK

Abstract
      Electronically conducting polymers have recently been considered
as electrode materials for various electrochemical devices, primarily
batteries and electronic displays. However, in the case of the polymers
yet studied, e.g. polyacetylene and polypyrrole, little regard has
been paid to the ionic conductivity which, according to ambipolar
diffusion theory, is required for mass transport of the electroactive
species from the surface to the bulk of the electrode. In fact,
experiments on polyacetylene in conjunction with solid polymer
electrolytes have indicated that the charge-discharge rate will be
limited by the ionic conductivity.
      A model of electrode operation involving electronic and ionic
resistance lines coupled by the charge storage capacitors can be used
to replace the more conventional diffusion treatment. This has the
advantage of facilitating the understanding of composite electrodes
in which the rather low redox storage capacitance of conducting polymers
may be supplemented by the addition of an inorganic insertion electrode.
Present formulations of composite electrodes, e.g. carbon black, PEO,
and $V_6O_{13}$ composite, suffer from an increased tortuosity of the
electronic conduction path due to the prescence of the ionic conductor,
and similarly that of the ionic path by the electronic conductor.
Therefore a mixed conductor with acceptable conductivities of both
carriers would be an advantage.
      Some preliminary experiments were described in which the polymer
design incorporates the structural elements required each for type
of conductivity, and the mass transport parameters are evaluated with
reference to the discharge rate of the polymer electrode in a solid
polymer electrolyte cell.

*L. Alcácer (ed.), Conducting Polymers, 103.*
© *1987 by D. Reidel Publishing Company.*

# POLYANILINE: SYNTHESIS AND CHARACTERIZATION OF THE EMERALDINE OXIDATION STATE BY ELEMENTAL ANALYSIS

A. G. MacDiarmid, J. C. Chiang, A. F. Richter and N. L. D. Somasiri, Department of Chemistry, University of Pennsylvania, Philadelphia, Pennsylvania, 19104-6323 (U.S.A.)
A. J. Epstein
Department of Physics, The Ohio State University, Columbus, Ohio, 43210, (U.S.A.)

ABSTRACT. Detailed experimental procedures are given for the chemical synthesis from aniline of analytically pure emeraldine hydrochloride, a highly conducting polymer derived from the emeraldine oxidation state of polyaniline, which contains equal numbers of oxidized and reduced repeat units, the non-protonated base form of which has the composition,

$[(-\langle\rangle-\overset{H}{N}-\langle\rangle-\overset{H}{N}-)(-\langle\rangle-N=\langle\rangle=N-)]_x$. In the as-synthesized polymer, ~ 42% of the nitrogen atoms are protonated i.e. "doped". Treatment of this material with 1.0M aqueous HCl gives by elemental analysis, the most highly conducting (metallic) form of the emeraldine oxidation state of polyaniline in which 50% of the nitrogen atoms are protonated. Experimental details are given for converting the as-formed emeraldine hydrochloride to analytically pure emeraldine base. The conductivities of samples of emeraldine base protonated by aqueous HCl to various extents as determined by elemental analysis are reported. Electrochemical studies involving the emeraldine base are consistent with its having a composition very close to the proposed composition involving equal numbers of oxidized and reduced repeat units.

## 1. INTRODUCTION

"Polyaniline" has been described in many papers [1-6] during the past 100 years or so. It has been reported as existing in various, usually ill-defined forms such as "aniline black", "emeraldine", "nigraniline", etc., synthesized by the oxidative chemical or electrochemical polymerization of aniline. Certain of these materials have been shown to have an unexpectedly high conductivity [2-6].

We believe that a series of polyanilines having the general formula

(in the base form), $\left[\left(\langle\rangle-\overset{H}{N}-\langle\rangle-\overset{H}{N}-\right)_y\left(\langle\rangle-N=\langle\rangle=N-\right)_{1-y}\right]_x$ , containing reduced repeat

units, $\langle\rangle-\overset{H}{N}-\langle\rangle-\overset{H}{N}-$ , and oxidized repeat units, $\langle\rangle-N=\langle\rangle=N-$ , can in principle be synthesized [6]. When $0 < y < 1$ they are poly(paraphenylene-amineimines) in which the oxidation state of the polymer continuously

L. Alcácer (ed.), Conducting Polymers, 105–120.

increases with decreasing value of y. The fully reduced poly(paraphenyl-
eneamine), "leucoemeraldine", $(-\langle\bigcirc\rangle-\overset{H}{N}-)_{2x}$ , corresponds to a value of y = 1;
the fully oxidized, poly(paraphenyleneimine), "pernigraniline",
$(=\langle\bigcirc\rangle=N-\langle\bigcirc\rangle-N=)_x$ , corresponds to a value of y = 0 and the 50% oxidized
poly(paraphenyleneamineimine), "emeraldine", $[(-\langle\bigcirc\rangle-\overset{H}{N}-\langle\bigcirc\rangle-\overset{H}{N}-)(\langle\bigcirc\rangle-N=\langle\bigcirc\rangle=N-)]_x$ ,
corresponds to a value of y = 0.5. The terms "leucoemeraldine", "emeral-
dine" and "pernigraniline" refer to different oxidation states of polyaniline. Each oxidation state can, in principle, exist in the form of
its base or as its protonated form (salt) by treatment of the base with
an acid [6].

Various chemical [1,2, 4-6] and electrochemical [3,4] procedures have
been used to synthesize polyaniline by the oxidative polymerization of
aniline. In no case has any complete elemental analysis been given for
emeraldine base or for any of its salts although from chemical reduction
studies of the base [1] it would appear that it has probably been obtained
in the pure state. Leucoemeraldine base is the only polyaniline in the
above series of compounds for which an elemental analysis has been
reported [1].

The present study was carried out for the purpose of finding an
experimental method for (i) synthesizing analytically pure emeraldine
base (ii) confirming that the product does indeed have the emeradine
oxidation state (iii) synthesizing analytically pure emeraldine hydrochlorides having a variety of different degrees of protonation (iv) determining the relationship between the degree of protonation and the conductivity of the salt.


2.   EXPERIMENTAL

2.1. Chemical Synthesis of Emeraldine Hydrochloride ( ~ 42% Protonation)
     from Aniline

Ammonium peroxydisulfate, $(NH_4)_2S_2O_8$ (11.5 g, 0.0504 mole) was dissolved in 200 ml of 1M HCl which had been precooled to 1°C. Aniline
(20.0 ml, 0.219 mole) was dissolved in 300 ml of 1M HCl which had been
precooled to 1°C. The aniline solution was placed in a 750 ml Erlenmeyer
flask with a magnetic stirring bar and the container was placed in an
ice bath on a magnetic stirring plate.

The $(NH_4)_2S_2O_8$ solution was added to the aniline solution, with
constant stirring, over a period of 1 minute. The solution was then
stirred in an ice bath for ~ 1.5 hours during which time the temperature
remained below 5°C. Three to 5 minutes after the reactants were mixed,
the solution started to take on a blue-green tint and then became intense
blue-green with a coppery glint as a precipitate formed. The coppery
glint was less pronounced after ~ 1 hour.

After ~ 1.5 hours, the precipitate was collected on a Buchner funnel
(diameter 7.5 cm) using a water aspirator. The precipitate cake was
washed portionwise (60 ml/portions) with 1M HCl until the initially pale
violet filtrate became colorless. The liquid level was constantly

adjusted so that it remained above the top of the precipitate. This prevented cracking of the precipitate cake, which would result in inefficient washing of the precipitate. A minimum of 500 ml of 1M HCl was used. This "as-made"precipitate is polyaniline in the incompletely protonated emeraldine hydrochloride form.

After the above washing, the precipitate remained under suction for ~ 10 minutes until significant cracking of the moist filter cake occurred. It was then transferred on the filter paper to a vacuum desiccator and dried under dynamic vacuum for ~ 2 hours. The large surface area of the product so exposed facilitated rapid removal of water. The semi-dry precipitate was then pulverized with a mortar and pestle and dried further on the vacuum line under dynamic vacuum at room temperature for ~ 12 hours. It was then heated under dynamic vacuum at ~ 85°C for 3 hours. Analysis [7]. Found: C, 67.53; H, 4.86; N, 13.25; Cl, 14.02% (Total C + H + N + Cl = 99.66%) Calc. for $C_{24}H_{18}N_4 \cdot 1.67HCl$ C, 68.09, H, 4.68; N, 13.24; Cl, 13.99%. Since one $H^+$ will accompany each $Cl^-$ which reacts with nitrogen in the polyaniline, the molar ratio of Cl to N as determined by elemental analysis will give the percent protonation of the polymer. Since in this analysis, (moles Cl)/(moles N) = 0.418, the percent protonation, i.e. percent "doping", is 41.8%. Since we believe that only the imine nitrogen atoms are protonated under the conditions used in this study, the maximum protonation, i.e. maximum "doping" level obtainable is 50%.

## 2.2. Conversion of As-synthesized Emeraldine Hydrochloride ( ~ 42% Protonation) to the Fully Protonated ( ~ 50%) Hydrochloride

The moist precipitate cake obtained by the 10 minute suction treatment in the Buchner funnel as described in Section 2.1, was equilibrated with constant stirring with 500 ml of 1M HCl for ~ 15 hours. The powder was collected on a Buchner funner (diameter 7.5 cm) and was washed with 500 ml of 1M HCl in 60 ml portions using similar precautions to those described in the previous section to avoid cracking of the filter cake. The powder was dried under suction on the Buchner funnel for ~ 10 minutes and was then transferred on the filter paper to a vacuum desiccator and was pumped for 4 hours. In all cases except sample 6 (Table I), the semidry precipitate was pulverized by mortar and pestle and was dried under dynamic vacuum for ~ 48 hours at room temperature. Sample 6 after pulverizing, was dried under dynamic vacuum for ~ 12 hours at room temperature and then at ~ 85°C for 3 hours under dynamic vacuum. The molar Cl/N ratios given in Table I are calculated directly from experimental nitrogen and chlorine elemental analyses. The mean value for the ratio is 0.499; the median value for the ratio is 0.489; the average deviation from the mean is ± 0.020. Hence the percent protonation ("doping" level), i.e., the percent of the total number of nitrogen

atoms in each repeat unit, $[(-\langle\!\!\!\!\rangle-\overset{H}{N}-\langle\!\!\!\!\rangle-\overset{H}{N}-)(-\langle\!\!\!\!\rangle-N=\langle\!\!\!\!\rangle=N-)]_x$ , which are protonated is 49.9 ± 2.0%.

Considerable difficulty was experienced at first with the elemental analyses [7] apparently due to incomplete combustion of the polymer. They were improved considerably by mixing a $V_2O_5$ catalyst with the polymer during combustion and by using a longer combustion time. Even so, it can be seen that the carbon analyses are usually somewhat low.

TABLE I    Analytical Data [7] for Emeraldine Hydrochloride ( ~ 50% Protonation) Obtained by Treatment of Chemically Synthesized Emeraldine Hydrochloride Powder ( ~ 42% Protonation) with 1M HCl

| Sample | Formula[a] | Cl/N[b] | % C | % H | % N | % Cl | Total % |
|--------|-----------|---------|-----|-----|-----|------|---------|
| 1 | $C_{24}H_{18}N_4 \cdot 1.95HCl$ | 0.489 | 63.76 | 5.17 | 12.79 | 15.84 | 97.56(Exptl.) |
|   |                                | 0.488 | 66.48 | 4.64 | 12.93 | 15.95 | 100.00(Calc.) |
| 2 | $C_{24}H_{18}N_4 \cdot 1.93HCl$ | 0.483 | 64.58 | 5.16 | 12.96 | 15.85 | 98.55(Exptl.) |
|   |                                | 0.482 | 66.60 | 4.64 | 12.95 | 15.81 | 100.00(Calc.) |
| 3 | $C_{24}H_{18}N_4 \cdot 2.10HCl$ | 0.536 | 64.92 | 4.55 | 11.96 | 16.22 | 97.65(Exptl.) |
|   |                                | 0.525 | 65.66 | 4.62 | 12.76 | 16.96 | 100.00(Calc.) |
| 4 | $C_{24}H_{18}N_4 \cdot 2.06HCl$ | 0.516 | 63.59 | 4.92 | 12.38 | 16.15 | 97.04(Exptl.) |
|   |                                | 0.515 | 65.88 | 4.62 | 12.81 | 16.69 | 100.00(Calc.) |
| 5 | $C_{24}H_{18}N_4 \cdot 2.05HCl$ | 0.514 | 64.07 | 4.45 | 12.56 | 16.32 | 97.40(Exptl.) |
|   |                                | 0.515 | 65.88 | 4.62 | 12.81 | 16.69 | 100.00(Calc.) |
| 6 | $C_{24}H_{18}N_4 \cdot 1.91HCl$ | 0.479 | 65.01 | 5.20 | 13.11 | 15.90 | 99.22(Exptl.) |
|   |                                | 0.480 | 66.68 | 4.61 | 12.96 | 15.75 | 100.00(Calc.) |
| 7[c] | $C_{24}H_{18}N_4 \cdot 1.91HCl$ | 0.477 |     |     | 12.26 | 14.81 | (Exptl.) |
|   |                                | 0.480 |     |     | 12.96 | 15.75 | (Calc.) |

(a)    The formula given is that which best fits the analytical data.

(b)    The first Cl/N ratio listed in each set was calculated directly from the experimental Cl and N analyses. The data so obtained was used in constructing the graph given in Figure 2. The second Cl/N ratio is calculated from the calculated empirical formula.

(c)    As a check on the reliability of the Cl and N analyses, the experimental molar Cl/N ratio was multiplied by four in order to obtain the number of Cl atoms per four N atoms present

in each $C_{24}H_{18}N_4$, $[(-\bigcirc-\overset{H}{N}-\bigcirc-\overset{H}{N}-)(-\bigcirc-N=\bigcirc=N-)]_x$, formula unit. From the empirical composition so obtained, a calculated value for the weight percent N and weight percent Cl was obtained.

## 2.3. Synthesis of Emeraldine Base

The moist emeraldine hydrochloride precipitate cake obtained after the 10 minute suction treatment in the Buchner funnel described in Section 2.1, was suspended with constant stirring in 500 ml 0.1M $NH_4OH$ solution. If, after 10 minutes, the pH of the suspended liquid was < 8, 1.0M $NH_4OH$ was added dropwise to bring the pH up to ~ 8. The suspension was stirred for ~ 15 hours. The powder was collected on a Buchner funnel (diameter 7.5 cm) and was washed with 500 ml of 0.1M $NH_4OH$ in 60 ml portions, precautions to avoid cracking of the filter cake being taken as described in Section 2.1. The powder was resuspended in an additional 500 ml of 0.1M $NH_4OH$ and was stirred for 1 hour, collected on a Buchner funnel and washed with 500 ml of 0.1M $NH_4OH$ in 60 ml portions. The powder was partially dried under suction on a Buchner funnel for ~ 10 minutes. The moist emeraldine base powder was then transferred on the filter paper to a desiccator and was dried under dynamic vacuum for ~ 4 hours. It was then pulverized by mortar and pestle and was dried further as described in Table II.

Similar analytical difficluties to those experienced with emeraldine hydrochloride were also encountered. It should be stressed that the emeraldine base and especially the emeraldine hydrochloride salts absorb water readily from the atmosphere. Organic solvents such as THF, DMSO, pyridine, $CH_3CN$ [5], 80% acetic acid [1], 60% formic acid [1], etc., can be used to extract varying amounts of lower molecular weight polymers from the emeraldine base synthesized as described above. Such solutions have a blue-green color. Extraction is continued until the solvent is colorless. Room temperature THF for example dissolves ~ 20% by weight of emeraldine base whereas DMSO dissolves ~ 80% [8] of the polymer. Our studies are all based on the convenient reference solvent, ~ 0.1M $NH_4OH$, which was colorless after being used to wash emeraldine base powder. The polymer so obtained may be extracted further by an appropriate organic solvent chosen according to the amount of lower molecular weight material which one wishes to extract.

TABLE II. Analytical Data [7] for Chemically Synthesized Emeraldine Base Powder, $C_{24}H_{18}N_4$, i.e [(-⟨◯⟩-Ṅ-⟨◯⟩-Ṅ-)(-⟨◯⟩-N=⟨◯⟩=N-)]$_x$ .

| Sample | % C | % H | % N | % Cl | Total % | Drying Conditions[a] |
|--------|-----|-----|-----|------|---------|---------------------|
| 1 | 78.72 | 5.28 | 15.60 | – | 99.60(Exptl.) | 1 |
|   | 79.53 | 5.01 | 15.46 | 0 | 100.00(Calc.) | |
| 2 | 78.26 | 5.06 | 15.90 | – | 99.22(Exptl.) | 2 |
|   | 79.53 | 5.01 | 15.46 | 0 | 100.00(Calc.) | |
| 3 | 78.29 | 5.00 | 15.05 | – | 98.34(Exptl.) | 2 |
|   | 79.53 | 5.01 | 15.46 | 0 | 100.00(Calc.) | |
| 4 | 79.54 | 5.07 | 14.06 | – | 98.67(Exptl.) | 2 |
|   | 79.53 | 5.01 | 15.46 | 0 | 100.00(Calc.) | |
| 5 | 79.25 | 4.95 | 14.43 | – | 98.63(Exptl.) | 2 |
|   | 79.53 | 5.01 | 15.46 | 0 | 100.00(Calc.) | |
| 6 | 80.21 | 5.02 | 15.21 | <0.4 | 100.44(Exptl.) | 3 |
|   | 79.53 | 5.01 | 15.46 | 0 | 100.00(Calc.) | |
| 7 | 78.79 | 4.95 | 15.83 | <0.4 | 99.57(Exptl.) | 3 |
|   | 79.53 | 5.01 | 15.46 | 0 | 100.00(Calc.) | |
| 8 | 78.34 | 5.74 | 14.39 | <0.4 | 98.47(Exptl.) | 2 |
|   | 79.53 | 5.01 | 15.46 | 0 | 100.00(Calc.) | |
| 9 | 78.35 | 5.11 | 15.91 | <0.4 | 99.37(Exptl.) | 2 |
|   | 79.53 | 5.01 | 15.46 | 0 | 100.00(Calc.) | |

(a) Drying Conditions (1) dynamic vacuum – 72 hours at room temperature

(2) dynamic vacuum – 48 hours at room temperature

(3) dynamic vacuum – 24 hours at room temperature, then 4 hours at 100°C

## 2.4. Alternate Methods of Synthesis of Emeraldine Base

Three 350 mg portions of emeraldine hydrochloride powder (Sample 6, Table I) were placed in separate 750 ml Erlenmeyer flasks together with magnetic stirring bars. They were then converted to emeraldine base by three different methods. An 0.1M $NH_4OH$ solution (500 ml) was added to the first flask; 500 ml of a solution of 0.1M $NH_4OH$ in a 50/50 volume mixture of $H_2O/CH_3OH$ was added to the second flask and 500 ml of a solution of 0.5M KOH in a 50/50 volume mixture of $H_2O/CH_3OH$ was added to the third flask. The above mixture were allowed to equilibrate with constant stirring for 16 hours in air.

The emeraldine base powders were collected individually on Buchner funnels (5 cm). The powder from the first flask was washed with 800 ml of water. The powders from the second and third flasks were each washed with 800 ml of a 50/50 volume mixture of $H_2O/CH_3OH$ and excess liquid was removed under suction on the Buchner funnels during ~ 10 minutes. Each filter paper and the adhering precipitate were transferred to a vacuum desiccator and were partially dried under dynamic vacuum for 2 hours.

The powders were pulverized using a mortar and pestle and were placed individually in 750 ml Erlenmeyer flasks containing 500 ml of a 0.010M HCl solution (Fisher Scientific Co.). The powders were equilibrated with the acid for 48 hours with constant stirring in air. The pH of each solution was measured after 48 hours (see Table III). The powders were collected in Buchner funnels and were partially dried under suction on a Buchner funnel for ~ 10 minutes. The moist emeraldine base powders were then transferred on the filter paper to a vacuum desiccator and were dried under dynamic vacuum for 2 hours and were then pulverized by mortar and pestle. They were then further dried under dynamic vacuum for ~ 48 hours. Elemental analyses for chlorine and nitrogen and four-probe conductivities were determined, (see Table III).

The compositions of the emeraldine hydrochloride synthesized from emeraldine base which was obtained by treatment of emeraldine hydrochloride by the different methods described above all fall within the expected range of values as can be seen from Figure 2. The conductivities also fall within the expected range (Figure 1 in [9]). Hence the above alternate methods of converting as-synthesized emeraldine hydrochloride to emeraldine base do not affect the nature of the emeraldine hydrochloride which can be synthesized from the base prepared by these alternate methods.

## 2.5. Synthesis of Emeraldine Hydrochloride (Protonation Levels < ~ 40%)

Emeraldine hydrochloride with protonation levels < ~ 40% was synthesized by equilibrating emeraldine base powder, ( ~ 250 mg) with stirring, for 55 hours in a large excess ( ~ 500 ml) of HCl solution of appropriate pH. As can be seen from Figure 1, it appears that equilibrium was completed well within this time. The powder was filtered and dried without washing. Analytical data and drying conditions are given in Table IV.

The experimental and calculated values for the chlorine and nitrogen analyses are, in general in reasonable agreement. The experimental molar

Cl/N ratios given in Table IV together with the experimental molar Cl/N ratios in Tables I and III are used in constructing the curve in Figure 2.

TABLE III.   Properties of Emeraldine Hydrochloride Obtained from Emeraldine Base Synthesized by Different Methods

Elemental Analyses [7]

| Method of Synthesis of Emeraldine Base | Final pH After Equilibrium in 0.01M HCl | | Cl/N[b] | % N | % Cl | $\sigma$ (S/cm)[a] |
|---|---|---|---|---|---|---|
| 0.1M NH$_4$OH in H$_2$O | 2.11 | (Exptl.) | 0.337 | 12.33 | 10.54 | 9.0 x 10$^{-2}$ |
| | | (Calc.) | 0.338 | 13.61 | 11.63 | |
| 0.1M NH$_4$OH in 50/50 volume H$_2$O/CH$_3$OH | 2.09 | (Exptl.) | 0.375 | 12.22 | 11.64 | 26 x 10$^{-2}$ |
| | | (Calc.) | 0.375 | 13.44 | 12.75 | |
| 0.5M KOH in 50/50 volume H$_2$O/CH$_3$OH | 2.06 | (Exptl.) | 0.344 | 11.89 | 10.39 | 9.5 x 10$^{-2}$ |
| | | (Calc.) | 0.345 | 13.58 | 11.85 | |

(a) Four probe conductivity on compressed pellet
(b) The first Cl/N ratio listed in each set was calculated directly from the experimental Cl and N analyses.  The data so obtained was used in constructing the graph given in Figure 2.  The second Cl/N ratio is calculated from the calculated empirical fromula.  As a check on the reliability of the Cl and N anaylses, the experimental molar Cl/N ratio was multiplied by four in order to obtain the number of Cl atoms per four N atoms present in each $C_{24}H_{18}N_4$,

$[(-\langle\bigcirc\rangle-\overset{H}{\underset{}{N}}-\langle\bigcirc\rangle-\overset{H}{\underset{}{N}}-)(-\langle\bigcirc\rangle-N=\langle\bigcirc\rangle=N-)]_X$, formula unit.  From the empirical composition so obtained, a calculated value for the weight percent N and weight percent Cl was obtained.

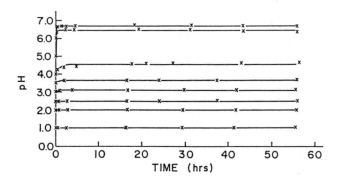

Figure 1.   Change in pH when emeraldine base powder is equilibrated with aqueous HCl solutions having various selected initial pH values.

TABLE IV.  Analytical Data [7] for Chemically Synthesized Emeraldine
Hydrochloride Salts (Protonation Level $< \sim 40\%$)[a]

| Sample | Formula | Final Equilibration pH | Cl/N[b] | % C | % H | % N | % Cl | Total% | |
|--------|---------|------------------------|---------|-----|-----|-----|------|--------|--|
| 1 | $C_{24}H_{18}N_4$ | 6.65 | 0.0008 | | | 14.49 | 0.03[c] | | (Exptl.) |
| | | | 0 | | | 15.46 | 0 | | (Calc.) |
| 2 | $C_{24}H_{18}N_4$ | 6.33 | 0.0003 | | | 14.37 | 0.01[c] | | (Exptl.) |
| | | | 0 | | | 15.46 | 0 | | (Calc.) |
| 3 | $C_{24}H_{18}N_4 \cdot 0.04HCl$ | 6.00 | <0.011 | | | 15.01 | <0.4 | | (Exptl.) |
| | | | 0.010 | | | 15.40 | 0.39 | | (Calc.) |
| 4 | $C_{24}H_{18}N_4 \cdot 0.01HCl$ | 4.60 | 0.0014 | | | 14.14 | 0.05[c] | | (Exptl.) |
| | | | 0.0026 | | | 15.45 | 0.10 | | (Calc.) |
| 5 | $C_{24}H_{18}N_4 \cdot 0.06HCl$ | 4.00 | <0.0141 | | | 14.05 | <0.5 | | (Exptl.) |
| | | | 0.0149 | | | 15.37 | 0.58 | | (Calc.) |
| 6 | $C_{24}H_{18}N_4 \cdot 0.11HCl$ | 3.64 | 0.0283 | | | 14.78 | 1.06 | | (Exptl.) |
| | | | 0.0274 | | | 15.29 | 1.06 | | (Calc.) |
| 7 | $C_{24}H_{18}N_4 \cdot 0.09HCl$ | 3.60 | 0.0232 | | | 14.51 | 0.85 | | (Exptl.) |
| | | | 0.0224 | | | 15.33 | 0.87 | | (Calc.) |
| 8 | $C_{24}H_{18}N_4 \cdot 0.40HCl$ | 3.43 | 0.0995 | 74.33 | 4.75 | 14.60 | 3.68 | 97.36 | (Exptl.) |
| | | | 0.1000 | 76.46 | 4.92 | 14.86 | 3.76 | 100.00 | (Calc.) |
| 9 | $C_{24}H_{18}N_4 \cdot 0.06HCl$ | 3.18 | <0.0141 | | | 12.50 | <0.5 | | (Exptl.) |
| | | | 0.0149 | | | 15.37 | 0.58 | | (Calc.) |
| 10 | $C_{24}H_{18}N_4 \cdot 0.12HCl$ | 3.12 | 0.0295 | | | 13.91 | 1.04 | | (Exptl.) |
| | | | 0.0300 | | | 15.28 | 1.16 | | (Calc.) |
| 11 | $C_{24}H_{18}N_4 \cdot 0.84HCl$ | 2.91 | 0.210 | 73.44 | 5.16 | 13.40 | 7.13 | 99.13 | (Exptl.) |
| | | | 0.210 | 73.33 | 4.83 | 14.26 | 7.58 | 100.00 | (Calc.) |
| 12 | $C_{24}H_{18}N_4 \cdot 0.78HCl$ | 2.51 | 0.194 | | | 14.04 | 6.89 | | (Exptl.) |
| | | | 0.195 | | | 14.34 | 7.07 | | (Calc.) |
| 13 | $C_{24}H_{18}N_4 \cdot 0.74HCl$ | 2.50 | 0.185 | | | 13.47 | 6.29 | | (Exptl.) |
| | | | 0.185 | | | 14.39 | 6.74 | | (Calc.) |
| 14 | $C_{24}H_{18}N_4 \cdot 1.00HCl$ | 2.43 | 0.250 | 72.00 | 4.86 | 13.14 | 8.33 | 98.33 | (Exptl.) |
| | | | 0.250 | 72.26 | 4.80 | 14.05 | 8.89 | 100.00 | (Calc.) |
| 15 | $C_{24}H_{18}N_4 \cdot 1.29HCl$ | 2.18 | 0.322 | | | 13.63 | 11.12 | | (Exptl.) |
| | | | 0.322 | | | 13.69 | 11.17 | | (Calc.) |
| 16 | $C_{24}H_{18}N_4 \cdot 1.11HCl$ | 2.16 | 0.278 | | | 13.42 | 9.44 | | (Exptl.) |
| | | | 0.278 | | | 13.91 | 9.77 | | (Calc.) |
| 17 | $C_{24}H_{18}N_4 \cdot 1.20HCl$ | 2.15 | 0.300 | | | 13.36 | 10.15 | | (Exptl.) |
| | | | 0.300 | | | 13.80 | 10.47 | | (Calc.) |
| 18 | $C_{24}H_{18}N_4 \cdot 1.31HCl$ | 2.15 | 0.328 | | | 13.59 | 11.29 | | (Exptl.) |
| | | | 0.328 | | | 13.66 | 11.32 | | (Calc.) |
| 19 | $C_{24}H_{18}N_4 \cdot 1.35HCl$ | 2.11 | 0.338 | | | 12.33 | 10.54 | | (Exptl.) |
| | | | 0.338 | | | 13.61 | 11.63 | | (Calc.) |
| 20 | $C_{24}H_{18}N_4 \cdot 1.32HCl$ | 2.10 | 0.329 | | | 13.50 | 11.23 | | (Exptl.) |
| | | | 0.330 | | | 13.65 | 11.40 | | (Calc.) |
| 21 | $C_{24}H_{18}N_4 \cdot 1.34HCl$ | 2.08 | 0.335 | | | 13.44 | 11.39 | | (Exptl.) |
| | | | 0.325 | | | 13.63 | 11.22 | | (Calc.) |
| 22 | $C_{24}H_{18}N_4 \cdot 1.31HCl$ | 2.00 | 0.327 | | | 12.43 | 10.29 | | (Exptl.) |
| | | | 0.328 | | | 13.66 | 11.32 | | (Calc.) |

TABLE IV Cont'd

| Sample | Formula | Final Equilibration pH | Cl/N[b] | % C | % H | % N | % Cl | Total % |
|--------|---------|------------------------|---------|-----|-----|-----|------|---------|
| 23 | $C_{24}H_{18}N_4 \cdot 1.13HCl$ | 2.00 | 0.283 | | | 12.45 | 8.90 | (Exptl.) |
|    |                                 |      | 0.282 | | | 13.88 | 9.92 | (Calc.) |
| 24 | $C_{24}H_{18}N_4 \cdot 1.44HCl$ | 1.49 | 0.359 | | | 13.94 | 12.66 | (Exptl.) |
|    |                                 |      | 0.360 | | | 13.51 | 12.30 | (Calc.) |
| 25 | $C_{24}H_{18}N_4 \cdot 1.37HCl$ | 1.48 | 0.342 | | | 12.99 | 11.24 | (Exptl.) |
|    |                                 |      | 0.343 | | | 13.59 | 11.78 | (Calc.) |
| 26 | $C_{24}H_{18}N_4 \cdot 1.50HCl$ | 1.40 | 0.374 | | | 13.40 | 12.69 | (Exptl.) |
|    |                                 |      | 0.381 | | | 13.44 | 12.95 | (Calc.) |
| 27 | $C_{24}H_{18}N_4 \cdot 1.76HCl$ | 1.00 | 0.440 | | | 12.50 | 13.94 | (Exptl.) |
|    |                                 |      | 0.440 | | | 13.14 | 14.63 | (Calc.) |

(a) All samples except sample 8, 11 and 14 were dried on the vacuum line under dynamic vacuum at room temperature for 48 hours. Samples 8, 11 and 14 were dried under dynamic vacuum for 12 hours followed by 3 hours at 85°C under dynamic vacuum.

(b) The first Cl/N ratio listed in each set was calculated directly from the experimental Cl and N analyses. The data so obtained was used in constructing the graph given in Figure 2. The second Cl/N ratio is calculated from the calculated empirical formula. As a check on the reliability of the Cl and N analyses, the experimental molar Cl/N ratio was multiplied by four in order to obtain the number of Cl atoms per four N atoms present in each

$C_{24}H_{18}N_4$, $[(-\langle\bigcirc\rangle-\overset{H}{N}-\langle\bigcirc\rangle-\overset{H}{N}+\langle\bigcirc\rangle-N=\langle\bigcirc\rangle=N-)]_x$ , formula unit. From the empirical composition so obtained, a calculated value for the weight percent N and weight percent Cl was obtained.

(c) Chlorine analyses for trace amounts of chlorine specially requested.

2.6. Equilibration of Emeraldine Hydrochloride and Emeraldine Base in Aqueous HCl of the Same pH

Moist, freshly prepared, chemically synthesized emeraldine hydrochloride ( ~ 50% protonation) ~ 0.3 g dry polymer, synthesized as described in Section 2.2, was divided into two equal portions. The first portion was held under argon atmosphere. The second portion was suspended with constant stirring in 500 ml of 0.2M $NH_4OH$ under an argon atmosphere for ~ 16 hours to convert it to emeraldine base as described in Section 2.3. It was collected while still in a moist-pasty form on a Buchner funnel under an argon blanket using an inverted funnel. Both powder samples were then placed in separate flasks containing 400 ml of 0.010M HCl (pH = 2.0) (Fisher Scientific Co.). The samples were stirred in air for ~ 2.5 hours. The pH of the solution containing the emeraldine hydrochloride was 1.97; the pH of the solution containing the emeraldine base was 2.20. The powders were collected by filtration and the filtrates were combined to give 800 ml of total solution. The powders were resuspended in 400 ml portions of the combined solution and stirred for 24 hours. The above procedure was repeated (4 times) until the pH of both

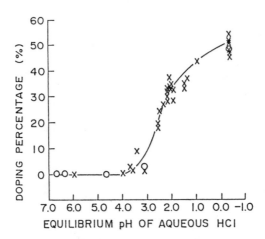

Figure 2. Relationship between doping percentage of emeraldine base and
equilibrium pH of HCl "doping" solution. (o = early data)

solutions were identical after 24 hours of stirring. The final pH of
both solutions was 2.065 ± 0.005. Both powders were then collected in
Buchner funnels.

    A small amount of each sample of the wet pasty powder was smeared
on a piece of glass filter paper (0.25 cm x 0.25 cm) which was then
crimped inside a Pt mesh. The potential of the powder was measured (vs.
SCE) in the final equilibration solution (pH of 2.065). The initial
potentials of both samples were identical within experimental error,
that obtained from emeraldine base being 0.355 V and that obtained from
emeraldine hydrochloride ( ~ 50% protonation) being 0.353 V (vs. SCE).
After a few seconds the potential of each sample began to increase slowly
as expected since subsequent studies have shown that in the presence of
platinum catalyst the emeraldine oxidation state of polyaniline is oxidized slightly by air in acid solutions. [10]

    The emeraldine oxidation state of polyaniline ( ~ 50% proton-
ation) (sample 5, Table I) had an immediate potential of 0.356 V (vs.
SCE) in the form of a pellet or powder encased in Pt gauze in 1M HCl [10]
(pH = - 0.2)[11]. A similar potential was observed when the polymer was
in contact with mercury as the current collector. This potential of
0.356 V may therefore be used to characterize the emeraldine oxidation
state of polyaniline. The above results show that the potential of the
emeraldine oxidation state is insensitive to the pH of the electrolyte
between pH = 2 and pH = - 0.2. This is consistent with a previous observation that the potential of an oxidation state of polyaniline, which is
somewhat more oxidized (potential = 0.438 V vs. SCE at pH = - 0.2) [10,
12] than the emeraldine oxidation state, is relatively insensitive to
the pH of the electrolyte in this same pH range.

    The observation that the potential of the emeraldine base and the
50% protonated emeraldine hydrochloride are identical when equilibrated
in solution of the same pH is consistent with, but does not prove that

they both ultimately attain the same degree of protonation. The inclusive
nature of this study is a result of the insensitivity of the potential of
the polymer, (which is a function of its degree of protonation [12]), to
the pH of the electrolyte in the pH range 2 to - 0.2.

## 2.7. Electrochemical Determination of Oxidation State of Chemically Synthesized Emeraldine Base Powder

As shown by Table V elemental analyses are not sufficiently accurate
to distinguish clearly between different oxidation states of polyaniline,
such as leucoemeraldine base and emeraldine base which differ by only two
hydrogen atoms per repeat unit consisting of four $(C_6H_4)N$ groups. Hence
the following electrochemical method was employed to determine the oxidation state of the analytically pure emeraldine base.

TABLE V.  Percent Weight Composition of Different Oxidation
State of Polyaniline

|  | % C | % H | % N |
|---|---|---|---|
| Leucoemeraldine Base | 79.09 | 5.53 | 15.38 |
| Emeraldine Base | 79.53 | 5.01 | 15.46 |
| Pernigraniline Base | 79.98 | 4.47 | 15.55 |

The electrochemical cell used consisted of a circular disc of Pt
$(1.5 \text{ cm}^2)$ to which a Pt wire was spot welded. This electrode was placed
inside the flat surface of a plastic vial cap (2.2 cm diameter) with the
Pt wire pointing upwards, perpendicular to the Pt disc. A Pyrex glass
tube (3 cm long, 1.0 cm internal diameter; $0.8 \text{ cm}^2$) was placed on the Pt
disc. The tube was then clamped and molten paraffin wax was poured into
the outside hollow area between the tube and the wall of the vial cap
until the wax level reached the top of the cap. The clamp was removed
after 15 minutes at which time the Pt disc was attached firmly to the
bottom of the glass tube by the paraffin wax with the Pt wire protruding
vertically through the paraffin wax seal.

The apparatus was then transferred to an argon-filled glove bag
which was purged three times to ensure that all air was removed. The
remainder of the cell construction was carried out under argon. A mixture
of 7.0 mg of emeraldine base (sample 7, Table II) and ~ 1.4 mg of carbon
black were placed on a glazed weighing paper. The mixture was then poured
down the glass tube so that it rested on the exposed Pt surface. Two
circular glass filter papers, each ~ $0.8 \text{ cm}^2$, were then placed on top of
the emeraldine powder base which was itself resting on the Pt disc.

A circular Zn electrode $(0.75 \text{ cm}^2, 0.25 \text{ cm thick})$, having a vertical

~ 5 cm "stem" was cut from a zinc sheet. The "stem" served as the connecting lead to the electrode. It was lightly amalgamated by dipping into 1.0M HgCl$_2$ solution for ~ 3 seconds, and was rinsed in distilled water. It was then placed in the cell so that circular flat Zn surface was in good contact with the glass filter paper. The electrode was then mechanically clamped firmly in place under downward pressure.

Approximately 2 ml of 1.0M ZnCl$_2$ solution (pH ~ 4) degassed previously by passing argon through the solution for one hour, was then poured down the glass tube. About 0.5 cm layer of molten paraffin wax was finally poured on to the top of the electrolyte. All experiments were carried out under argon inside a glove bag.

The cell described above exhibited an immediate preequilibrium initial open circuit voltage of 1.18 V (vs. Zn) i.e. 0.18 V (vs. SCE). The emeraldine base was then reduced to leucoemeraldine base during 97 minutes at a constant current of 0.75 mA/cm$^2$. The area referred to is th 0.75 cm$^2$ of Pt disc inside the circular glass tube. This reduction step was terminated when the final cell potential reached 0.5 V (vs. Zn) i.e. −0.5 V (vs. SCE). The relationship between the cell potential and the coulombs in the reduction reaction is shown in Figure 3. As can seen, no

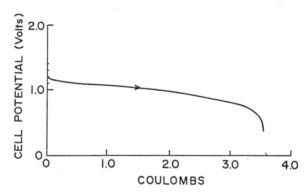

Figure 3. Relationship between the change in potential and charge passed during the reduction of chemically-synthesized emeraldine base in 1.0M aqueous ZnCl$_2$ solution.

significant reduction occurred after the cell potential reached 0.65 V (vs. Zn) when the cell potential began to drop sharply. The total charge liberated when the cell potential reached 0.65 V was 3.50 C. The calculated charge required for complete reduction of 7.0 mg of emeraldine base to leucoemeraldine base according to Equation (1) is 3.73 C.

$$[(-\langle\rangle-\overset{H}{N}-\langle\rangle-\overset{H}{N}-)(-\langle\rangle-N=\langle\rangle=N-)]_X + (2x)H^+ + (2x)e$$

$$\downarrow \tag{1}$$

$$[(-\langle\rangle-\overset{H}{N}-\langle\rangle-\overset{H}{N}-)(-\langle\rangle-\overset{H}{N}-\langle\rangle-\overset{H}{N}-)]_X$$

If the error in weighing the 7.0 mg of emeraldine base is ± 0.2 mg, then the oxidation state of the analytically pure emeraldine base used in ths study can be expressed by the formula:

$$[(\text{—}\bigcirc\text{—}\overset{H}{\underset{N}{}}\text{—}\bigcirc\text{—}\overset{H}{\underset{N}{}}\text{—})_{0.531 \pm 0.013}(\text{—}\bigcirc\text{—}N=\bigcirc=N\text{—})_{0.469 \pm 0.013}]_x$$ . Hence the polyaniline appears to have an oxidation state close to that of emeraldine,

$$[(\text{—}\bigcirc\text{—}\overset{H}{\underset{N}{}}\text{—}\bigcirc\text{—}\overset{H}{\underset{N}{}}\text{—})(\text{—}\bigcirc\text{—}N=\bigcirc=N\text{—})]_x$$ .

## 3.   RESULTS AND DISCUSSION

The chemical oxidative polymerization of aniline in aqueous HCl solution by ammonium peroxydisulphate, $(NH_4)_2S_2O_8$ is a modification of previously-published methods [2]. Elemental analyses for hydrogen are not sufficiently accurate to determine the extent of protonation, ("doping" level) of the polymer. However, it can be safely assumed that each chloride ion found in the polymer will be accompanied by one proton. Hence the overall extent of protonation can be assessed accurately from an elemental chlorine analysis. It is assumed that only nitrogen atoms are protonated.

In the as-synthesized emeraldine hydrochloride, elemental analyses show that ~ 42% of the total number of nitrogen atoms present in the polymer are protonated, no distinction being made between amine and imine nitrogen atoms. This material has a conductivity of ~ 1 s/cm. The as-synthesized emeraldine hydrochloride can be further protonated by equilibration with aqueous 1.0M HCl approaching a maximum protonation value of ~ 50% in 1.0M HCl (pH = - 0.2) [16]. In this acid, as can be seen from Table I, the average protonation level obtained by elemental analyses of samples from seven separate syntheses is 49.9 ± 2.0%. Thus 50% of all the nitrogen atoms in the polymer are protonated under these experimental conditions. The conductivity increases with protonation level reaching a maximum of ~ 5 s/cm at 50% protonation when dried as described [6,9]. The conductivity does not change significantly even when protonated in 10M aqueous HCl [6], (pH = - 3.59) [11]. We believe that protonation occurs in strong HCl according to equation (2). The

$$[(\text{—}\bigcirc\text{—}\overset{H}{\underset{N}{}}\text{—}\bigcirc\text{—}\overset{H}{\underset{N}{}}\text{—})(\text{—}\bigcirc\text{—}N=\bigcirc=N\text{—})]_x$$

$$\downarrow \quad (2x)HCl$$

$$[(\text{—}\bigcirc\text{—}\overset{H}{\underset{N}{}}\text{—}\bigcirc\text{—}\overset{H}{\underset{N}{}}\text{—})(\text{—}\bigcirc\text{—}\overset{H}{\underset{\overset{N}{\underset{Cl^-}{}}}{}}=\bigcirc=\overset{H}{\underset{\overset{N}{\underset{Cl^-}{}}}{}}\text{—})]_x \qquad (2)$$

$$\downarrow \quad \text{internal redox reaction}$$

$$[(\text{—}\bigcirc\text{—}\overset{H}{\underset{\overset{N}{\underset{Cl^-}{}}}{}}\text{—}\bigcirc\text{—}\overset{H}{\underset{N}{}}\text{—})(\text{—}\bigcirc\text{—}\overset{H}{\underset{N}{}}\text{—}\bigcirc\text{—}\overset{H}{\underset{\overset{N}{\underset{Cl^-}{}}}{}}\text{—})]_x$$

protonated imine may not necessarily be an intermediate species in the
spontaneous internal redox reaction. The metallic emeraldine hydrochloride can therefore be regarded as a polysemiquinone radical cation
[9]. It has been shown to exhibit strong Pauli magnetic susceptibility
and is believed to involve a polaronic conduction band [9,13].

The as-synthesized emeraldine hydrochloride can be readily converted
to analytically pure emeraldine base, an insulator, by treatment with
aqueous alkali. The three slightly different methods presented apparently
all give equally pure material.

Equilibration of emeraldine base with aqueous HCl of different
concentrations results, as can be seen from Figure 2, in increased levels
of protonation as the acidity of the HCl is increased. The acid can be
removed reversibly by washing with dilute aqueous base, or more slowly by
washing with water.

We believe it is very important to use a volatile acid when
synthesizing an emeraldine salt. If an aqueous solution of a non-volatile
acid such as $H_2SO_4$ is used, evaporation of the solution from the emeraldine sulfate powder could result in the formation of a thin layer of
concentrated $H_2SO_4$ on the surface of each powder particle. This may
react further with the polymer, but in any case it is certainly expected
to affect many of its properties. Removal of the excess acid by washing
with water will bring about some immediate deprotonation with consequent
loss of conductivity; similarly, washing with basic solvents such as
$CH_3CN$ might also be expected to result in some deprotonation.

It can be seen from Tables I, II, and IV that the experimental
elemental analysis for carbon usually tends to be a little low although
the improved analytical methods described in the experimental section
have greatly reduced the originally serious problem not infrequently
encountered when analyzing polymers of this type. It should be noted,
however, that both for the emeraldine hydrochlorides and for the emeraldine base, the sum of the weight percent compositions in most cases is
very close to 100% indicating that no extensive amounts of oxygen-
containing material such as water or quinoid, $\Longleftarrow$ o groups are present.

One of the more puzzling problems still remaining is that as we, [6]
and others [2,14] have noted, the conductivity of emeraldine salts seems
to be increased to varying extents by the presence of traces of water
vapor to which the polymer may be inadvertently or purposefully exposed.

It must be stressed that elemental analyses cannot distinguish
reliably between different oxidation states of polyaniline such as those
given in Table V. However, as shown in Section 2.7, slow electrochemical reduction gave a composition of

$[(-\langle\bigcirc\rangle-\overset{H}{N}-\langle\bigcirc\rangle-\overset{H}{N}-)_{0.531 \pm 0.013}(-\langle\bigcirc\rangle-N=\langle\bigcirc\rangle=N-)_{0.469 \pm 0.013}]_x$ for the analytically
pure chemically-synthesized emeraldine base used in the present study.
This composition is in reasonably good agreement with the $TiCl_3$ reduction
[1] of chemically synthesized emeraldine base to the leucoemeraldine

oxidation state, $[(-\langle\bigcirc\rangle-\overset{H}{N}-\langle\bigcirc\rangle-\overset{H}{N}-)(-\langle\bigcirc\rangle-N=\langle\bigcirc\rangle=N-)]_x$ . The results of this
chemical volumetric analytical study show that the oxidation state of

the emeraldine base employed corresponded to the composition,

$[(-\bigcirc-\overset{H}{N}-\bigcirc-\overset{H}{N}\text{+}]_{0.510}\text{+}\bigcirc-N=\bigcirc=N\text{-})_{0.490}]_x$ , which is very similar to that of

the emeraldine oxidation state, $[(-\bigcirc-\overset{H}{N}-\bigcirc-\overset{H}{N}\text{-})(\bigcirc-N=\bigcirc=N\text{-})]_x$ . It is
somewhat surprising and indeed fortuitous that the oxidation potential
of the $(NH_4)_2S_2O_8$ solution used in the synthesis is apparently appropriate
to give the form of polyaniline which corresponds almost exactly to that
of the emeraldine oxidation state. The above results also indicate that
capacitance effects are apparently negligible under the slow electrochemical redox conditions employed in this present study even though
there is a considerable body of published data which suggest that
capacitance effects (which are not related to any chemical redox reaction), may be very large when polyaniline and certain other conducting
polymers are electrochemically oxidized and reduced rapidly in cyclic
voltammetry studies [15,16] or even under considerably slower conditions
[16].

In order to avoid confusion in this rapidly expanding area of research involving the polyanilines it appears to be absolutely essential
that the exact conditions under which the synthesis of a given form of
polyaniline be described and that some type of analytical data be provided
to indicate the composition, oxidation state etc. of the polymer under
investigation.


4.   ACKNOWLEDGEMENT

These studies were supported by the University of Pennsylvania Materials
Research Laboratory through N.S.F. grant No. DMR-8216718 (J.C.C.,
N.L.D.S.), and by the Office of Naval Research (A.F.R.).


5.   REFERENCES

(1)   A. G. Green and A.E. Woodhead, J. Chem. Soc. 97 (1910) 2388. ibid.,
      101 (1912) 1117.
(2)   R. de Surville, M. Jozefowicz, J. H. Perichon and R. Buvet, Electro-
      chim. Acta, 13 (1968) 1451: M. Jozefowicz, J. H. Perichon, L. T.
      Yu and R. E. Buvet, U.K. Patent No. 1 216 549 (1970).
(3)   J. Langer, Solid State Commun., 26 (1978) 839; A. F. Diaz and J. A.
      Logan, J. Electroanal. Chem., 111 (1980) 111; G. Mengoli, M. T.
      Munari, P. Bianco and M. M. Musiani, J. Appl. Polym. Sci., 26
      (1981) 4247; R. Noufi, A. J. Nozik, J. White and L. F. Warren, J.
      Electrochem. Soc., 129 (1982) 2261; T. Kobayashi, H. Yoneyama and
      H. Tamura, J. Electroanal. Chem., 161 (1984) 419; 177 (1984)
      281; 177 (1984) 293; T. Ohsaka, Y. Ohnuki, N. Oyama, G. Katagiri
      and K. Kamisako, J. Electroanal. Chem., 161 (1984) 399; A. Kitani,
      J. Izumi, J. Yano, Y. Hiromoto and K. Sasaki, Bull. Chem. Soc.
      (Japan), 57 (1984) 2254; E. M. Genies, A. A. Syed and C. Tsintavis,
      Mol. Cryst. Liq. Cryst., 121 (1985) 181; D. W. DeBerry, J. Electro-
      chem. Soc., 132 (1985) 1022; C. M. Carlin, L. J. Kepley and A. J.

Bard, J. Electrochem. Soc., 132 (1985) 353; M Kaneko and H. Nakamura, J. Chem. Soc., Chem. Commun. (1985) 346 E. W. Paul, A. J. Ricco and M. S. Wrighton, J. Phys. Chem., 89 (1985) 1441.

(4)    A. G. MacDiarmid, J. C. Chiang, M. Halpern, W. S. Huang, S. L. Mu, N. L. D. Somasiri, W. Wu and S. I. Yaniger, Mol. Cryst. Liq. Cryst., 121 (1985) 173; A. G. MacDiarmid, S. L. Mu, N. L. D. Somasiri and W. Wu, Mol. Cryst. Liq. Cryst., 121 (1985) 187; W. R. Salaneck, B. Liedberg, O. Inganas, R. Erlandsson, I. Lundstrom, A. G. MacDiarmid, M. Halpern and N. L. D. Somasiri, Mol. Cryst. Liq. Cryst., 121 (1985) 191; A. G. MacDiarmid, J. C. Chiang, W. S. Huang, B. D. Humphrey and N. L. D. Somasiri, Mol. Cryst. Liq. Cryst., 125 (1985) 309.

(5)    J. P. Travers, J. Chroboezek, F. Devreux, F. Genoud. M. Nechtschein, A. Syed, E. M. Genies and C. Tsintavis, Mol. Cryst. Liq. Cryst., 121 (1985) 195.

(6)    J. C. Chiang and A.G. MacDiarmid, Synth. Met., 13 (1986) 193.

(7)    Schwarzkopf Microanalytical Laboratory, Woodside, N.Y., 11377 (USA)

(8)    M. Angelopoulos and A. G. MacDiarmid, unpublished observations, (1986).

(9)    A. G. MacDiarmid, J. C. Chiang, A. F. Richter and A. J. Epstein, Synth. Met., in press (1986).

(10)   A. G. MacDiarmid, J. C. Chiang, A. F. Richter, unpublished observation (1986).

(11)   C. H. Rochester, Acidity Function, Academic Press, New York, 1970, p. 39.

(12)   W. S. Huang and A. G. MacDiarmid, submitted to J. Am. Chem. Soc., (1986).

(13)   A. J. Epstein, J. M. Ginder, F. Zuo, R. W. Bigelow, H. S. Woo, D. B. Tanner, A. F. Richter, W. S. Huang and A. G. MacDiarmid, Synth. Met., in press (1986).

(14)   M. Nechschein, C. Santier, J. P. Travers, J. Chroboczek, A. Alix and M. Ripert, Synth. Met., in press (1986).

(15)   T. Kobayashi, H. Yoneyama and H. Tamura, J. Electroanal. Chem., 177 (1984) 281.

(16)   N. Mermilliod, J. Tanguy, M. Hoclet and A. A. Syed, Synth. Met., in press (1986).

# ARE SEMICONDUCTING POLYMERS POLYMERIC SEMICONDUCTORS?: POLYANILINE AS AN EXAMPLE OF "CONDUCTING POLYMERS"

Arthur J. Epstein
Department of Physics and Department of Chemistry
The Ohio State University
Columbus, Ohio 43210-1106

John M. Ginder
Department of Physics
The Ohio State University
Columbus, Ohio 43210-1106

Alan F. Richter and Alan G. MacDiarmid
Department of Chemistry
University of Pennsylvania
Philadelphia, Pennsylvania 19104

ABSTRACT. The general concepts of quasi-one-dimensional conducting polymers are introduced including the role of band theory, electron-phonon interactions, the Peierls ground state, and commensurability. The dominant defect states present upon doping, solitons, polarons and bipolarons, are discussed. Application of these concepts to polyaniline is made with emphasis on the mechanism for the insulator-to-metal transition.

## 1. INTRODUCTION

Conducting polymers have now been studied for nearly a decade since the report [1] of the achievement of a highly conducting "metallic" state upon doping of polyacetylene, $(CH)_x$, with acceptors such as iodine and $AsF_5$ and donors such as sodium. This report led to vigorous activity in exploration of the physics and chemistry of the phenomena associated with these systems. A review of the full extent of theoretical development as well as synthetic and experimental work is beyond the scope of a short article. Therefore, only a brief introduction to the physical concepts as applied to a quasi-one-dimensional system is given below, followed by application to a conducting polymer of current interest, polyaniline.

121

L. Alcácer (ed.), Conducting Polymers, 121–140.
© 1987 by D. Reidel Publishing Company.

## 2.    BACKGROUND

In order to develop the specific picture, we begin with an ideal case [2] of a chain of N atoms evenly spaced a distance $a$ apart, as in Figure 1.

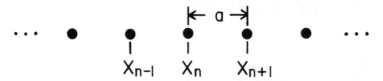

Figure 1.    Uniform chain of atoms with separation $a$.

Each atom is assumed to have an atomic orbital or wavefunction $\phi$ at each site n, $\phi_n$. Using a linear combination of atomic orbitals or tight binding approximation, these wavefunctions can be added to make a wavefunction of the entire system:

$$\Psi_k = \frac{1}{\sqrt{N}} \sum_n e^{ikan} \phi_n \qquad (1)$$

The term $e^{ikan}$ is a phase factor in the addition of the atomic orbitals. This combination of atomic orbitals leads to a spreading out of the energy level of the individual orbital from $E_0$ to a band states of width $W = 4t$, where t is the transfer (or $\beta$) integral. The index k of the phase factor can be used to index the new orbitals of the entire chain, which have an energy given by

$$E(k) = E_o - 2t\cos(ka) \qquad (2)$$

Figure 2(a) illustrates the spreading out of the energy level into a band of states.

The energies now run from $(E_0-2t)$ to $(E_0+2t)$ and the index k runs from a value of $-\pi/a$ to $+\pi/a$. According to the Pauli exclusion principle, each orbital $\phi_n$ can accommodate 0, 1, or 2 electrons; thus the total band of states can accommodate a maximum of 2N electrons. The Fermi energy $E_F$ denotes the energy of the highest occupied state in the band while the Fermi wavevector $k_F$ denotes the momentum index for the highest occupied state. If the energy band is completely empty or completely full, the material is an insulator; if it is partly occupied, it is a metal. If there is a small energy gap between a filled band and an empty band, it is a semiconductor.

The density of states $N(E) = (dE/dk)^{-1}$ expresses the number of states per unit energy within the band. For the tight binding band of Figure 2(a), the number of states per unit energy changes smoothly near the center of the band. However, it diverges at the bottom and the top of the band because the energy bands change very slowly with k, coming in with zero slope at $k=0$

and $k = \pm \pi/a$. These singularities in the density of states are indicated in Figure 2(b). For the tight binding energy band, the density of states, $N(E)$, is

$$N(E) = \frac{N/\pi}{[4t^2 - (E - E_0)^2]^{\frac{1}{2}}} \qquad (3)$$

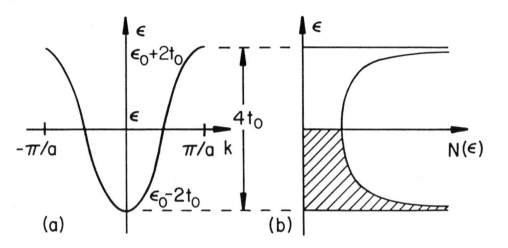

Figure 2.     (a) Energy-wavevector relation and (b) electron density of states for a one-dimensional tight-binding band. Shading denotes occupied states for a half-filled band.

This model can be applied to the ideal *trans*-polyacetylene chain assuming sp2 hybridization of the carbon $p_x$ and $p_y$ orbitals with the carbon 2s electron to form the fully occupied $\sigma$ bands of the polymer. The $p_z$ orbital extends perpendicular to the plane of the polymer chain. If the carbon atoms were uniformly spaced, the ideal tight-binding band with transfer integral $t_0$ given by the matrix element between adjacent $p_z$ orbitals would apply. As there is only one $p_z$ electron per carbon atom, the resulting $\pi$ band would be only one-half filled with band structure and density of states shown in Figure 2.

The transfer integral, t, depends upon the separation between adjacent sites [3]. If the location of each site is $x_n$ and the average site distance is $a$, then the deviation from uniform spacing $u_n$ is given by

$$u_n = x_n - na \qquad (4)$$

For $u_{n+1} - u_n < 0$, i.e., closer approach of adjacent carbon atoms, the transfer integral increases. Similarly for $u_{n+1} - u_n > 0$, that is for increased separation of adjacent atoms, the transfer integral decreases.    This is illustrated schematically in Figure 3.

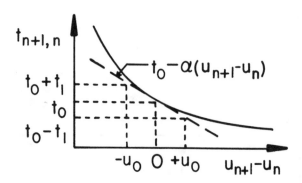

Figure 3.      Schematic dependence of the near-neighbor transfer integral, $t = t_{n+1,n}$, on atomic separation (after Reference 3).

For a dimerized lattice, the sites move alternately closer together and further apart as in Figure 4.  This leads to two types of transfer integrals, $t_0 + t_1$ for the increased transfer integral between sites that are closer together, and $t_0 - t_1$ for the reduced transfer integral for sites further apart. This variation of the transfer integral from that of the uniform chain can be approximated (Figure 3) to be linearly proportional to the deviation from uniform spacing or

$$t = t_o - a(u_{n+1} - u_n) \qquad\qquad (5)$$

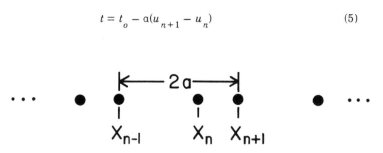

Figure 4.      Dimerized chain of atoms.

Figure 5 illustrates the effect of this new periodicity along the chain of atoms on the energy bands and density states.  A (Peierls [4]) gap is seen to open up in the energy band at $E_0$ with an energy gap equal to $2\Delta_c$.  This energy gap converts the formerly metallic, half-filled energy band system to one with a gap at the Fermi energy and, hence, a semiconductor.

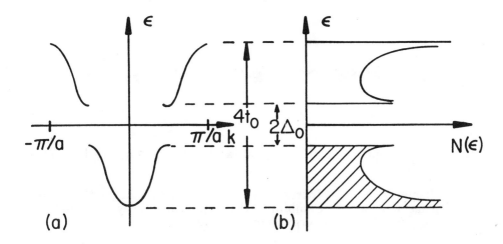

Figure 5.    (a)Energy-wavevector relation and (b) electron density of states
for a dimerized chain.  Shading denotes occupied states for a
half-filled band (after Reference 3).

The magnitude of the energy gap can be related [3] to the displacement
of the atom positions from uniformity along the chain.  Assuming $u_n = (-1)^n u$,
then

$$E_g = 2\Delta_c = 8a|u| \tag{6}$$

The energy gap, $E_g$, thus is directly proportional to the magnitude of the
displacement from uniformity along the chain with a constant of
proportionality $a$ ($a$ relates the transfer integral change to the change in
lattice spacing).  Here $a$ is the electron-phonon (or lattice displacement)
coupling constant.

Examination of Figure 5(b) shows that dimerization leads to a net
lowering of the occupied energy levels.  The electronic energy gain is related
to the magnitude of the displacement from uniform position of the lattice and
is proportional to $-u^2 \ln|u|$ for small u.  The presence of the electronic energy
gain alone would lead to collapse of the lattice.  However, displacement from
uniformity of the chain also costs an elastic energy.  Representing the elastic
energy along the chain (due to the σ bonds) as a set of springs connecting the
mass of the atoms, dimerization of the lattice leads to compression of some
springs and stretching of others, Figure 6.

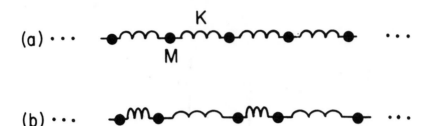

Figure 6.    Mass-spring idealizations of (a) uniform chain; (b) dimerized
             chain.

For small displacement, this increase in energy is proportional to the square
of the displacement or $u^2$. Combining the electronic energy gain and the
elastic energy loss, the total energy change is given by

$$E(u) = au^2 - bu^2 \ln|u|$$                                          (7)

where a and b are constants. Figure 7 illustrates this variation of energy per
repeat unit. A minimum is seen for a displacement of either $+u_0$ or $-u_0$,
representing the dimerization of even and odd atoms together or odd and even
atoms together, the two phases of bond alternation being labeled the A phase
and B phase, respectively.

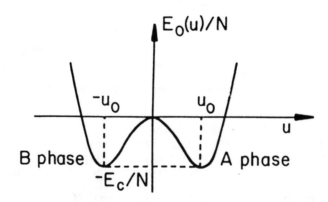

Figure 7.    Dependence of the total energy per repeat unit on the
             dimerization parameter u (after Reference 3).

The energy band picture and density of states, Figure 5, is now reminiscent of that of a semiconductor; indeed, for polyacetylene an energy gap of 1.4 eV is measured and a conductivity at room temperature of $10^{-5}$ (ohm-cm)$^{-1}$ is known [5]. These values are similar to those of the semiconductor silicon. Doping of polyacetylene with anions such as $I_3^-$ and $ClO_4^-$ and cations such as $Na^+$ leads to an increase in conductivity to nearly metallic values as illustrated in Figure 8.

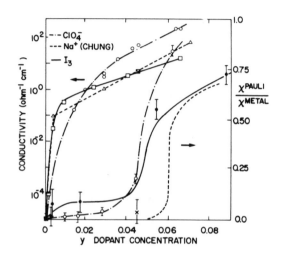

Figure 8.    Conductivity and Pauli susceptibility as a function of dopant concentration y in *trans*-(CH)$_x$(data for $I_3^-$, $C10_4^-$ and $Na^+$ are from Refs. 6-9).

This qualitative similarity to the behavior of semiconductors such as silicon upon doping was noted in early research. However, for silicon n or p type doping leads to the formation of holes in the valence band (HOMO) or electrons in the conduction band (LUMO) and a concomitant increase in the number of spins. Figure 8 illustrates the behavior for polyacetylene for doping levels as high as .08 charge carrier per carbon atom. At low doping levels, the magnetic susceptibility shows a very small or nearly zero spin concentration [6]. Beyond 5% doping, the spin susceptibility rapidly increases and becomes similar to that of the Pauli susceptibility of a metal with a band structure such as that in Figure 2(a). This unusual spin and charge relationship is indicative of the fact that though the energy bands and band gap are similar to those of inorganic semiconductors, the phenomena that occur upon doping are different.

The origins of the phenomenon of nonmagnetic charge carriers in polyacetylene were suggested in 1979 by three groups: Su, Schrieffer and Heeger [3], Michael Rice [10] and Serge Brazovskii [11]. The nucleation of a region of A phase within a polymer chain that is otherwise entirely B phase requires the formation of two boundary walls, a soliton S and an antisoliton S̄ between the two phases. As the energy per unit length of the polymer is the same whether in the A phase or B phase, the additional energy required for connecting between the A phase and the B phase (the creation energy per soliton) is $E_s = 2\Delta/\pi$ [12]. Associated with each soliton is an energy level at midgap, Figure 9. This energy level is singly occupied if the domain wall is neutral; it is doubly occupied and spinless if the domain wall is negatively charged and unoccupied and spinless if the domain wall is positively charged. In order to reduce the elastic energy, the domain wall is spread out over approximately 15 carbon atoms in polyacetylene [3]. As noted above, the cost of creation of each of the defect levels is $2\Delta/\pi$ (less than $\Delta$) but the cost of putting an electron on or taking an electron off this level is zero since it is at midgap. Hence, it is energetically favorable to create soliton defects for the storage of charge rather than to put the excess in either the conduction band or the valence band. The relevance of the soliton mechanism for charge storage in the regime up to 5% doping in polyacetylene has been well verified by many experiments [5], although the details of the models are modified for Coulomb interaction, interchain interactions, etc.

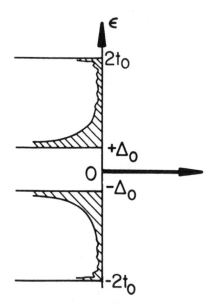

Figure 9.    Change in the electron density of states due to introduction of a soliton showing the midgap soliton state (after Reference 3).

It is useful to describe the periodicity of the lattice in terms of an "order parameter", $\Psi_n$. For the A phase $\Psi_n = (-1)^n u_n = u_0$ while for the B phase $\Psi_n = (-1)^{n+1} u_n = -u_0$. Figure 10 illustrates the variation of $\Psi_n$ for the A phase and B phase with distance along the chain, x, with $x_0$ the location where the polymer chain changes the phase of bond alternation. The soliton smoothly connects the A phase and B phase together (connecting $u_0$ to $-u_0$), while the antisoliton connects the B phase to the A phase, as shown in Figure 11. Solutions for the soliton and antisoliton order parameter are

$$\Psi_n = -u_0 tanh((x_n - x_0)/l)$$

$$\Psi_n = u_0 tanh((x_n - x_0)/l) \tag{8}$$

where $l$ is the soliton half-width ($l \sim 7a$ in $trans$-$(CH)_x$[3]).

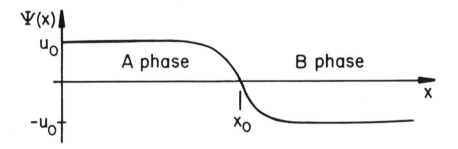

Figure 10.   Variation of the order parameter $\Psi(x)$ with position x. A soliton connects a region of A phase ($\Psi(x) = u_0$) to a region of B phase ($\Psi(x) = -u_0$).

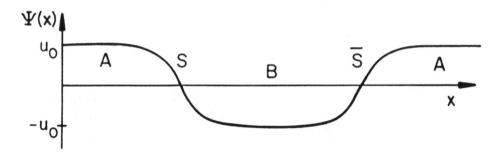

Figure 11.   $\Psi(x)$ for a region of B phase nucleated in A phase showing soliton S and antisoliton $\bar{S}$.

For an infinitely long isolated polyacetylene chain, the two phases of the bond alternation, A and B, are of the same energy. However, for many other polymers or  polymers in a crystalline structure, the two phases may not be equal in energy. For example, the B phase may have a larger energy per unit length than the A phase. If this occurs, then the soliton and antisoliton will not separate far upon their formation. Two soliton wave functions can be combined together to make new energy levels and new particles. For example, one can form a polaron by combining a soliton  and an antisoliton:

$\psi^{polaron} = u_0 - c[tanh((x-x_2)/l) - tanh((x-x_1)/l)]$ [13], as illustrated in Figure 12(a).

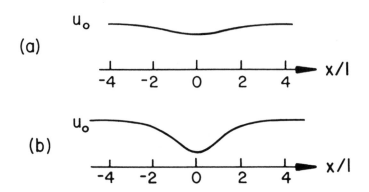

Figure 12.    (a) $\Psi(x)$ for a polaron and (b) $\Psi(x)$ for a bipolaron, using parameters appropriate for *cis*-polyacetylene (after Reference 14).

The polaron is thus a localized distortion of bonds within a uniform bond alternation pattern. If the soliton and the antisoliton comprising the polaron are both charged, the local bonding pattern can further distort, forming a bipolaron as indicated in Figure 12(b) [9] . A schematic energy level diagram is shown in Figure 13 where the energy levels of the soliton and antisoliton are combined together to form energy levels at $-\omega_1$ and $+\omega_1$. The occupation of those energy levels and the charge/spin relations are summarized in Table I. Using solitons as building blocks, we can form singly charged polarons with spin or doubly charged bipolarons without spin.

In sum, the energy levels of the individual atoms along the chain form a band of states. For example, the π band that results from the $p_z$ orbital of $(CH)_x$ would be partly filled and polyacetylene would be metallic. However, alternation of the bond distances leads to an energy gap in the polymer, converting the metallic polymer to a semiconductor. Because of the origin of the energy gap in the placement of the atoms, addition of charge leads to the formation of defect states such as solitons, polarons and bipolarons rather than electrons and holes in the conduction and valence bands respectively. Although not discussed in detail here, the soliton, polaron, or bipolaron energy levels may in turn overlap to form their own bands of states [15], these bands being located within the larger π-π* band gap formed by the dimerization of the carbon backbone. The addition of Coulomb interaction, interchain interaction and disorder are necessary for a more complete description of the phenomena in these systems.

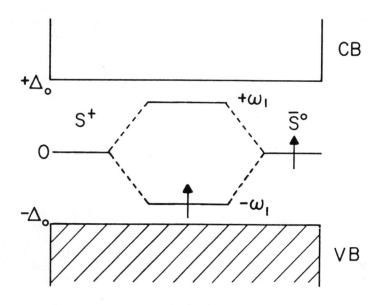

Figure 13.    Schematic hybridization of soliton energy levels at midgap to form polaron or bipolaron levels at $\pm\omega_1$. It is illustrated here for $S^+ + \bar{S}^0 \to P^+$ as for example in *trans*-polyacetylene.

TABLE I

| Defect type | Occupation | | Net charge | Spin |
|:---:|:---:|:---:|:---:|:---:|
| | $-\omega_1$ | $+\omega_1$ | | |
| $P^+$ | 1 | 0 | $+1$ | $\frac{1}{2}$ |
| $P^-$ | 2 | 1 | $-1$ | $\frac{1}{2}$ |
| $BP^{++}$ | 0 | 0 | $+2$ | 0 |
| $BP^{--}$ | 2 | 2 | $-2$ | 0 |

Occupation numbers, net charges, and spins for polarons (P) and bipolarons (BP).

The origin of the transition from a nonmagnetic highly conducting state to a magnetic conducting state at 5% doping in polyacetylene remains controversial. Several models have been proposed, one of which is that at high doping, the soliton energy levels overlap to form a band of states. This soliton band is completely empty if produced by p type doping or completely filled if created by n type doping. It has been proposed the presence of disorder in this system due to charged impurities at random interstitial sites and lack of perfect crystallinity lead to band tailing into the energy gaps, formation of a finite density of states, and Pauli susceptibility [16]. This phase disordering of the Peierls semiconductor is supported by numbers of experiments on p-type doped polyacetylene [17]. Alternatively it has been proposed that at high doping concentration, it is energetically more favorable for the soliton energy band to reorganize as polaron energy bands leading to a transition from a completely empty (filled) soliton band to a half filled lower (upper) polaron band for p(n) type doping [18]. This latter mechanism was in particular proposed[18] for n type doping of polyacetylene. A third mechanism, originally applied to polyparaphenylene, is the overlap of the bipolaron energy levels with the valence band to produce a band of states with a finite density of states at $E_F$ [19].

## 3.    APPLICATION OF CONCEPTS: POLYANILINE

There has been extensive successful application of the concepts introduced in the previous section to conducting polymers, especially polyacetylene. In recent years there has been increasing emphasis on the development of new conducting polymer systems such as polyparaphenylene, poly(1,6-heptadiyne), polythiophene and polypyrrole [5]. These systems are similar in that they become conducting through oxidation (removal of electrons from the polymers' π electron structure). An alternative type of conducting polymer is exemplified by the emeraldine form of polyaniline [20-29] as illustrated in Figure 14. This polymer consists of equal numbers of oxidized (Fig. 15(a)) and reduced (Fig. 15(b)) units. The relative ratio of the oxidized (concentration x) and reduced (concentration 1-x) units can be controlled from the completely oxidized to the completely reduced through electrochemical modification.

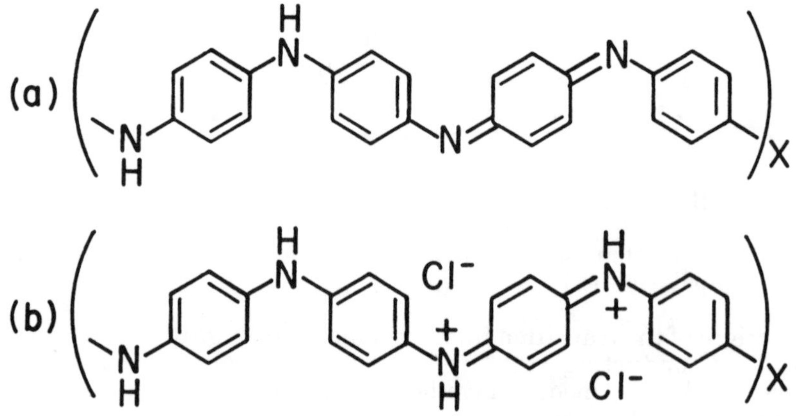

Figure 14.    (a) The idealized emeraldine base (EB) and (b) emeraldine salt (ES) forms of polyaniline.

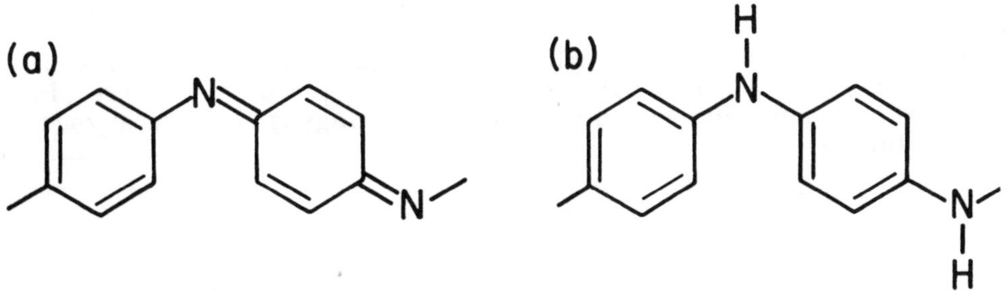

**(a)**  **(b)**

Figure 15.    (a) Oxidized, (b) reduced repeat units for polyaniline.

The emeraldine base form ($x = 0.5$), Figure 14(a) has a conductivity at room temperature of $\sim 10^{-10}$ S/cm. Upon treatment of the emeraldine base form with acids of varying pH, protons, $H^+$, are added to a fraction of the formerly unprotonated nitrogen sites. For example, treatment of emeraldine base (EB) with aqueous HCl of pH $\sim 0.0$ yields the highly conducting ($\sigma \sim 5$ S/cm) nearly completely protonated state (emeraldine salt, Figure    14(b)).    Thus, in contrast to the polymers listed above, the number of electrons on the polymer backbone is held constant while the number of protons is varied.    The conducting form of polyaniline can also be synthesized through the electrochemical oxidation of the leucoemeraldine (LE) form of the polymer which, before oxidation, is comprised of the reduced repeat units, Figure 15(b).

The chemical preparation of the emeraldine base form of polyaniline and its protonation to the salt form is described in MacDiarmid et al [20]. Materials were prepared both in the powder form and as films cast from solution. The magnetic susceptibility was measured via the Faraday technique [30] and electron spin resonance. Measurement of the temperature and protonation dependence of the magnetic susceptibility provides a direct probe of the development of a nonzero density of states at the Fermi energy and, hence, a metallic ground state [31]. The temperature dependent magnetic susceptibility, $\chi^M$, of selected samples of emeraldine as a function of protonation (chlorine to nitrogen ratio) is shown in Figure 16 [32,33]. The data were decomposed into contributions due to core diamagnetism, $\chi^{core}$, temperature dependent localized spins, $\chi^{curie}$, and Pauli susceptibility, $\chi^{Pauli}$:

$$\chi^M = \chi^{core} + \chi^{Curie} + \chi^{Pauli} \tag{9}$$

where $\chi^{core}$ is determined from the diamagnetic contributions of the polymer backbone plus those of any protons and counterions added to the polymer, $\chi^{Curie} = C/T$ where C is the Curie constant for noninteracting $s = \frac{1}{2}$ spins with $g = 2$ and $\chi^{Pauli} = \mu_B^2 N(E_F)$ where $N(E_F)$ is the density of states of both signs of spin at the Fermi energy. From the data for the emeraldine base, an experimental value for the core diamagnetism is obtained:

$$\chi_{EB}^{Core} = -106 \times 10^{-6} \, emu/mole - 2ring$$

This value was consistent with values calculated from Pascal constants for the individual components of the polymer with the proposed structure. In addition, approximately one Curie spin per 400 rings is observed at room temperature.

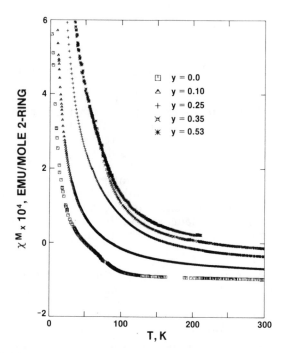

Figure 16.    Molar magnetic susceptibility $\chi^M$ versus temperature T for emeraldine samples of representative protonation levels.

Using Equation 9 together with the experimental $\chi_{EB}{}^{core}$ and $\chi_{Cl^-}{}^{core} = -25 \times 10^{-6}$ emu/mole-Cl⁻, the Pauli susceptibility for each of the samples as a function of protonation can be evaluated [33] by fitting the data for $T > 100K$ to Equation 1. For samples that are completely protonated (i.e., Cl/N = 0.5), $\chi^{Pauli} = \sim110 \times 10^{-6}$ emu/mole 2 ring. As each 2 ring repeat unit contains 14 heavy atoms (carbon and nitrogen), this value of $\chi^{Pauli}$ corresponds to $7.1 \times 10^{-6}$ emu/mole heavy atom. This compares with $3 \times 10^{-6}$ emu/mole C for polyacetylene at $\sim.06$ dopant per carbon atom. Hence the Pauli susceptibility for heavily protonated emeraldine is approximately $2\frac{1}{2}$ times that of that of polyacetylene doped with the same charge per heavy atom. Further, the number of Curie spins increases substantially upon protonation in contrast to the decrease observed upon doping of polyacetylene [9,30]. The Pauli susceptibility as a function of protonation is shown in Fig. 17. The roughly linear increase in $\chi^{Pauli}$ with concentration suggests phase segregation into fully protonated and unprotonated regimes, with the $\chi^{Pauli}$ for the metallic regime equal to ~ 110 x 10$^{-6}$emu/mole H$^+$. Figure 18(a) depicts the concept of full protonation with the positive charge at either side of the quinoid structure within the emeraldine polymer. The reversal of the phase of the bond alternation in proceeding along a polymer chain from a benzenoid ring to a quinoid ring and to the next benzenoid ring with a charge on either side of the quinoid is suggested to be a doubly positive charged bipolaron. This bipolaron entity has, as indicated in Table I, charge q = +2|e| but spin 0. The presence of a very sizable Pauli susceptibility together with a significant number of Curie spins implies that the bipolaron is unstable in protonated emeraldine despite the prediction of its relative stability within the continuum model [14].

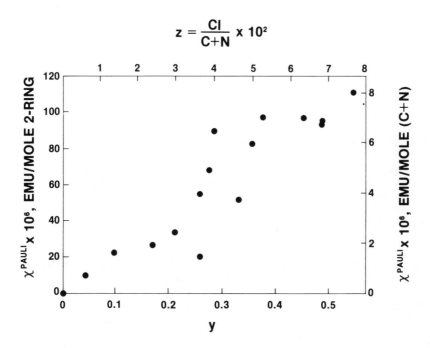

Figure 17. Pauli susceptibility $\chi^{Pauli}$ as a function of doping level y = Cl/N.

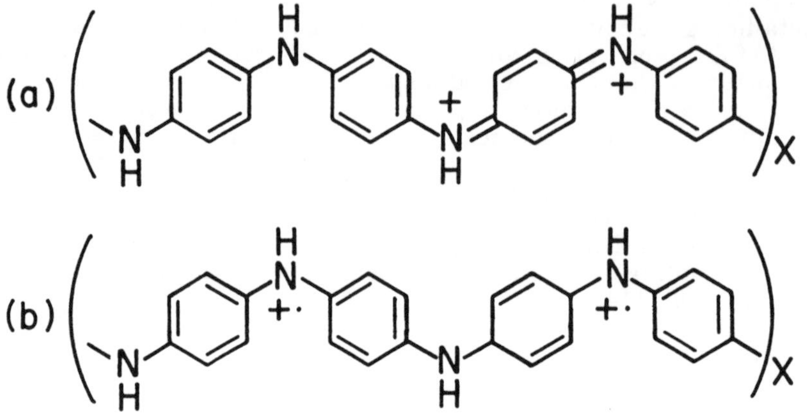

Figure 18.    Schematic (a) bipolaron lattice and (b) polaron lattice in the
emeraldine salt polymer.

The origin of the Pauli susceptibility and hence the finite density of
states at the Fermi energy, $N(E_F) = \chi^{Pauli}/\mu_B^2$, is important for
understanding polymers with nondegenerate ground states. The lack of
Drude behavior in the optical absorption as a function of frequency from the
infrared through visible [21] argues against the metallic state resulting
between overlap between a bipolaron band derived from overlap of the
bipolarons centered on the quinoid rings with the valence band of the
emeraldine as had been proposed [19] for polyparaphenylene. Similarly, the
high Pauli susceptibility and density of states argue against phase
disordering of the Peierls semiconductor as has been proposed [16] for p-type
doped polyacetylene as the dominant mechanism. The observation that the
Curie spin  density is much less than the number of protons added argues
against localized independent charged defects with spin.

It has been suggested [33] that the localized bipolarons are not the
stable ground state for the charges but rather there is a phase transition to a
polaron lattice. In this case there is a redox behavior transferring charge to
the quinoid group, converting it to an aromatic ring, resulting in two spins
being formed. The separation of the spins and charges by an average of two
nitrogen-nitrogen spacings leads to the formation of a polaron lattice, as is
schematically indicated in Fig. 18(b). In this case the polarons are shown
centered on the every other nitrogen. The transition from a bipolaron on
every fourth ring to a polaron centered on every second nitrogen is similar to
the soliton lattice to polaron lattice transition proposed [18] for n-doped
polyacetylene. Though a single bipolaron is energetically preferred over two
polarons [14], the Coulomb interaction between the charges within the
bipolaron together with the banding energy, dielectric screening and disorder
may drive the bipolaron to dissociate into two polarons.

The optical spectrum [21] shows the change from one absorption band for the emeraldine base to two new ones for the emeraldine salt. The excitation at 2eV in the emeraldine base has been identified as an exciton [21] formed by excitation of an electron from a quinoid ring (with a hole remaining in the quinoid ring) to the adjacent benzenoid rings [34]. The appearance of two new excitations at 1.5eV and 2.9eV for the protonated material are in agreement with the expectation of the two energy levels for polarons. However, polyaniline does not have a half-filled π band and so lacks charge conjugation symmetry; consequently, the two polaron energy levels may not form in the same manner as in polyacetylene [34-38]. Assuming [33] a one-dimensional tight binding model for the polaron band, the measured Pauli susceptibility predicts a polaron bandwidth $W_p = 0.37eV$ and a polaron decay length $\xi_0 = 0.4$ nitrogen-nitrogen spacings.

Electronic transport experiments support a picture of phase segregation into small metallic islands [21,39]. The conductivity of heavily protonated polyaniline is proportional to $\exp[-(T_0/T)^{\frac{1}{2}}]$ with $T_0$ a constant over the temperature range studied [21,39]. These and other studies of the electric field behavior of the conductivity [21,39] are in agreement with the charging energy limited tunnelling model [40] suggesting very small metal particles separated by insulating barriers.

## 4.    SUMMARY

Numerous new concepts have been developed to treat semiconducting polymers. The role of the electron-phonon interaction, one-dimensionality, Coulomb repulsion, and disorder are much more marked than they are in three-dimensional inorganic semiconductors such as silicon. Though the conductivity and band gap of the conducting polymers may be similar in magnitude to those of silicon, the detailed behaviors are quite different. Application of these models has been quite successful in developing a framework for polyacetylene including both its static and dynamic properties. We have presented here a proposed application of these concepts to polyemeraldine, which becomes highly conducting through protonation rather than through oxidation. Further modification of these concepts will be necessary to account for the presence of alternating NH and $C_6$ units in the polymer chain and the loss of charge conjugation symmetry.

# REFERENCES
Supported in part by the Office of Naval Research (JMG, AFR).

1.  H. Shirakawa, E.J. Louis, A.G. MacDiarmid, C.K. Chiang, and A.J. Heeger, *J.C.S. Chem. Comm.*, 578 (1977).

2.  J.S. Miller and A.J. Epstein, *Progress in Inorganic Chemistry* **20**, 1 (1976).

3.  W.P. Su, J.R. Schrieffer and A.J. Heeger, *Phys. Rev. Lett.* **42**, 1698 (1979); *Phys. Rev.* B **22**, 2099 (1980).

4.  R.E. Peierls, *Quantum Theory of Solids* (Clarendon, Oxford, 1955), p.108.

5.  See e.g. *Handbook of Conducting Polymers*, Vols. **1** and **2**, edited by T.A. Skotheim, (Marcel Dekker, Inc., New York (1986)).

6.  A.J. Epstein, R.W. Bigelow, H. Rommelman, H.W. Gibson, R.J. Weagley, A. Feldblum, D.B. Tanner, J.P. Pouget, J.C. Pouxviel, R. Comes, P. Robin and S. Kivelson, *Mol. Cryst. Liq. Cryst.* **117**, 147 (1985).

7.  A.J. Epstein, H. Rommelmann, R. Bigelow, H.W. Gibson, D.M. Hoffman and D.B. Tanner, *Phys. Rev. Lett.* **50**, 1866 (1983).

8.  A. Feldblum, R. Bigelow, H.W. Gibson, A.J. Epstein, and S. Kivelson, to be published.

9.  T.C. Chung, F. Moraes, J.D. Flood, and A.J. Heeger, *Phys. Rev.* B **29**, 2341 (1984).

10. M.J. Rice, *Phys. Lett.* **A71**, 152 (1979).

11. S.A. Brazovskii, *JETP Lett.* **28**, 656 (1978).

12. H. Takayama, Y.R. Lin-Liu, and K. Maki, *Phys. Rev.* B **21**, 2388 (1980).

13. D.K. Campbell, A.R. Bishop and K. Fesser, *Phys. Rev.* B **26**, 6862 (1982).

14. Y. Onodera, *Phys. Rev.* B **30**, 775 (1984).

15. B. Horovitz, *Phys. Rev. Lett.* **46**, 742 (1981).

16. E.J. Mele and M.J. Rice, *Phys. Rev.* B **23**, 5397 (1981).

17. X.Q. Yang, D.B. Tanner, M.J. Rice, H.W. Gibson, A. Feldblum and A.J. Epstein, *Solid State Commun.*, in press.

18. S. Kivelson and A.J. Heeger, *Phys. Rev. Lett.* **55**, 308 (1985).

19. J.L. Bredas, B. Thémans, J.G. Fripiat, J.M. Andre and R.R. Chance, *Phys. Rev.* B **29**, 6761 (1984).

20. J.C. Chiang and A.G. MacDiarmid, *Synth. Met.* **13**, 193 (1986).

21. A.J. Epstein, J.M. Ginder, F. Zuo, H.-S. Woo, D.B. Tanner, A.F. Richter, M. Angelopoulous, W.-S. Huang, and A.G. MacDiarmid, *Synth. Met.*, in press, and references, therein.

22. A.G. Green and A.E. Woodhead, *J. Chem. Soc.*, 2388 (1910).

23. R. de Surville, M. Josefowicz, L.T. Lu, J. Perichon and R. Buvet, *Electrochim. Acta* **13**, 1451 (1968).

24. A.G. MacDiarmid, J.-C. Chiang, M. Halpern, W.S. Huang, J.R. Krawcyzk, R.J. Mammone, S.L. Mu, N.L.D. Somasiri and W. Wu, *Polymer Preprints* **25**, 248 (1984).

25. P.M. McManus, S.C. Yang and R.J. Cushman, *J. Chem. Soc., Chem. Commun.*, 1556 (1985).

26. J.P. Travers, J. Chroboczek, F. Devreux, F. Genoud, M. Nechtschein, A. Syed, E.M. Genies and C. Tsintavis, *Mol. Cryst. Liq. Cryst.* **121**, 195 (1985).

27. E.M. Genies and C. Tsintavis, *J. Electronal. Chem.* **195**, 109 (1985).

28. W.R. Salaneck, I. Lundström, W.-S. Huang and A.G. MacDiarmid, *Synth. Met.* **13**, 291 (1986).

29. G.E. Wnek, *Polymer Prepr.* **27**, 277 (1986).

30. A.J. Epstein, H. Rommelman, M.A. Druy, A.J. Heeger and A.G. MacDiarmid, *Solid State Commun.* **38**, 683 (1981).

31. A.J. Epstein, J.M. Ginder, F. Zuo, R.W. Bigelow, H.-S. Woo, D.B. Tanner, A.F. Richter, W.-S. Huang and A.G. MacDiarmid, *Synth. Met*, in press.

32. A.G. MacDiarmid, J.C. Chiang, A.F. Richter, and A.J. Epstein, *Synth. Met.*, in press.

33. J.M. Ginder, A.F. Richter, A.G. MacDiarmid and A.J. Epstein, submitted.

34. C.B. Duke, E.M. Conwell and A. Paton, *Chem. Phys. Lett.*, submitted.

35. S. Stafstron and J.L. Bredas, *Synth. Met.* **14**, 297 (1986).

36. R.R. Chance, D.S. Boudreux, J.F. Wolf, L.W. Shacklette, R. Silbey, B. Themans, J.M. Andre and J.L. Brédas, *Synth. Met.* **15**, 105 (1986).

37. C.B. Duke, A. Paton, E.M. Conwell, W.R. Salaneck and I. Lundström, submitted.

38. S. Stafstrom, submitted.

39. F. Zuo, M. Angelopolous, A.G. MacDiarmid, and A.J. Epstein, to be published.

40. P. Sheng, B. Abeles and Y. Arie, *Phys. Rev. Lett.* **31**, 44 (1973).

# NEW ELECTRONICALLY CONDUCTING POLYMERS

F. Wudl
Institute for Polymers and Organic Solids
Department of Physics
University of California, Santa Barbara, California 93106

ABSTRACT. New conducting polyanilines consisting of alternating p-phenylenediamines with fluorene, biphenyl (benzidine), 4,4-diamino-E-stilbene or pyrene were prepared by an efficient polycondensation reaction. These materials show the typical pH-dependent properties of polyaniline. The fluorene and benzidine polymers are produced in their "leuco" (fully reduced form) and are Brönsted **base** dopable. The structure and properties of polyaniline were fully confirmed via the preparation of a completely characterized octamer.

## 1.0   INTRODUCTION

Within the recent past, the field of polymeric conductors has seen the emergence of soluble materials which could be cast into films[1] and the reinvestigation of polyaniline, particularly in reference to its unusual pH-dependent physical properties[2]. We have ongoing projects on soluble polythiophenes and polyaniline. In this paper we discuss our most recent work on poly-p-phenyleneamines, the first members of what appears to be a large polyaniline family.

### 1.1 Poly-p-phenyleneamines

The structure as well as the mechanism of Brönsted acid doping and the mechanism of electronic conductivity of PANI have been the subject of much

L. Alcácer (ed.), Conducting Polymers, 141–147.

debate in the very recent past[2]. We recently reported our results on the structure determination of PANI *via* synthesis[3]. That project provided us with a bonus; the preparation of new poly-p-phenyleneamines.

Ever since the first reported preparation of PANI over 140 years ago[4], its structure determination was hampered by insolubility. Also, the structure could not be inferred from its method of preparation because the first intermediate on the way to the polymer could couple nonregio - specifically at positions ortho or para to the nitrogen atom. In addition, a carbon-carbon bond forming reaction such as benzidine formation could not be ruled out *a priori*. The evidence gathered by Willstäter[4] strongly implicated a regular array of para-nitrogen-linked phenylene (**B**) and quinoneimine (**Q**) units. The color of the product was claimed to be dependent on the number of **Q**'s in the chain. Thus, "leuco emeraldine" (cream), "emeraldine" (blue), "nigraniline" (bluish-black with bronze luster), and "pernigraniline" (purple-brown) were names assigned to an assumed octamer with a backbone consisting of zero, two, three, and four **Q**'s, respectively. We were surprised to learn that the oldest method of structure proof, synthesis, had not been reported. We have shown that a material qualitatively identical to PANI can be prepared by a decarboxy - lative condensation reaction in which there could be **no benzidine** and **no ortho-coupled product** formation[3]. However, the number of **B**'s and **Q**'s in the blue form could not be ascertained directly from the polycondensation.

## 1.2 Phenyl-capped Ocataaniline (COA)

In order to test if indeed PANI was an octaaniline, as proposed by Willstäter and others[4] and resurrected by Wnek[5], and in order to be able to make an oligomer with a **known, specific number of quinoneimines**; we pre- pared, employing our condensation reaction mentioned above, a phenyl- capped octamer. We show below that three well defined oxidation states; leuco (white), per (magenta-violet), and emerald (blue) could be prepared and characterized.

Having established complete control over the chemistry of COA, we set out to gain a better understanding of the phenomena associated with pH- dependent physical changes in PANI, assuming that COA was a good model.

## 2.0 RESULTS AND DISCUSSION

### 2.1 Poly-p-phenyleneamines

Our synthesis of polyaniline was a modification of works of Honzl[6] and Moore[7] and involved the polycondensation of para-phenylenediamine with 2,5-dihydroxy-1,4-dihydroterephthalic acid (DHDHTA). The reaction afforded a blue-black product in a crude yield of 90%[8]. The achievement of such a high yield encouraged us to determine the scope of this synthesis for the preparation of a potentially large group of "polyanilines" (poly-p-phenyleneamines) resulting from the condensation of essentially any aromatic diamine and DHDHTA as shown below.

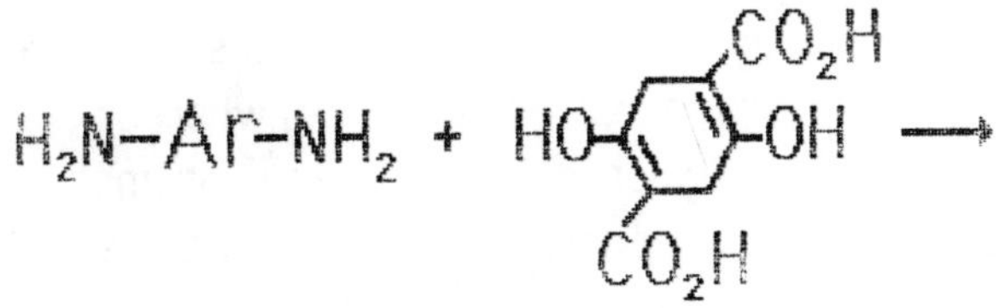

$$H_2N-Ar-NH_2 \;+\; HO\!-\!\!\langle\rangle\!-\!OH \;\longrightarrow$$

with $CO_2H$ (top) and $CO_2H$ (bottom) substituents

**1**

$$(\!-\!HN-Ar-NH\!-\!\langle\rangle\!-\!)_n \xrightarrow{\;[o]\;}$$

with $CO_2H$ (top) and $CO_2H$ (bottom) substituents

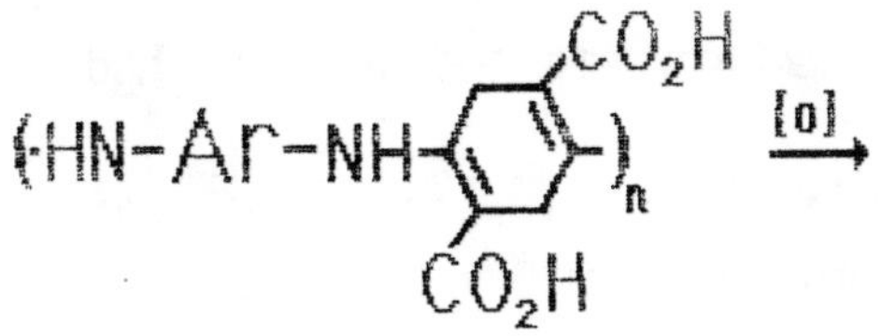

**2**

$$\text{(-HN-} \ \text{Ar-NH-} \underset{\overset{|}{CO_2H}}{\overset{CO_2H}{\bigcirc}} \text{)}_n$$

3

We are pleased to be able to report here preliminary results which indicate that benzidine, 4,4'-diamino-E-stilbene, 2,7-diaminofluorene (DAF) and 1,6-diaminopyrene all give deeply colored polymers with DHDHTA. The products exhibit the unique pH-dependent properties of PANI with some differences peculiar to each diamine. Much to our surprise, all diamines reacted at qualitatively different rates! For example, whereas p-phenylene-diamine produces a tan dihydropolymer with complete consumption of starting diamine (tlc) within 48 hr, under the same conditions, 2,7-di - aminofluorene and benzidine take *ca* 6 days to be converted completely to a brown dihydropolymer.

All polymers show the typical three PANI peaks in the 1600 - 1150 $cm^{-1}$ region of their infrared spectrum plus other expected peaks, e.g., C-H stretching vibrations at $2940cm^{-1}$ for the methylene hydrogens of the fluorenediamine polymer.

So far we have not examined these polymers in detail but have obtained some preliminary data mostly on the DAF, benzidine and diaminopyrene polymers. The former and the latter are blue, the other is dark grey, when in the neutral state. Solutions in DMF (N,N-dimethyl - formamide) which are light brown, turn green upon acidification with hydrogen chloride. The electronic spectra of the as-formed DAF and benzidine polymers in solution revealed that they were essentially in the leuco form and that they could be Brönsted **base** doped.

The as-formed DAF diaminopyrene and benzidine polymers have a two-probe compaction resistivity of $>10^7\Omega cm$. Exposure of the DAF solid to hydrogen chloride causes its resistivity to decrease to $10^3 - 10^4\Omega cm$, PANI and the diaminopyrene polymer under the same conditions (hydrate) have a resistivity of ~5 - 10 and $10^{-1}\Omega$cm, respectively. Just as in the case of PANI[3], the number of unpaired spins per repeat unit increases by *ca* two orders of magnitude upon acidification of the solid diaminopyrene polymer.

## 2.2 Phenyl-capped Octaaniline (COA)

When two molecular weights of an aromatic monoamine were allowed to condense with one molecular weight of DHDHTA, a p-phenylenediamine was produced. Thus, Ph-NH-Ph-NH-Ph-NH-Ph-NH$_2$ ("emeraldine base")[4] afforded "COA blue", a material with an unknown number of **Q**'s which could be reduced to COA (white, leuco) with phenylhydrazine.

This polyaniline oligomer differs from the actual structure for leuco polyaniline proposed ~80 years ago only in that a terminal -NH$_2$ was converted to -NHPh. The material is soluble to some extent in DMF but insoluble in most common organic solvents. In DMF solution, it is easily oxidized by air to the blue derivative but in the solid state it is oxidized only slowly. Spectroscopy of this blue material shows that it is identical in all respects to polyaniline. However, on the basis of solubililty properties, we believe that PANI is probably of higher molecular weight than COA. The two-probe compaction resistivity of the as-formed COA blue is $> 10^7 \Omega$cm. Hydrogen chloride treatment produces a green material with an increase of $\sim 10^2$ spins, as determined by ESR but a resistivity decrease to only $\sim 10^3 \Omega$cm. When the amount of **Q** in the backbone is increased, the resis - tivity of the green material drops to within that observed for PANI (see below).

Oxidation of COA with either PbO$_2$ or CrO$_3$, produces a dark brown - to - black colored solid which is *insoluble* in DMF but soluble to some extent in chloroform to afford a magenta-purple solution. This material could be shown to be the fully oxidized COA; i.e., TQI (phenyl-capped tetraquinone - imine). Surprisingly, the infrared spectrum of this material turned out to be qualitatively the same as that of COA in the 1600 - 1150cm$^{-1}$ region. The quantitative difference was in the relative intensities of the 1580 and 1490cm$^{-1}$ peaks; in COA, 1490 > 1580cm$^{-1}$ and *vice versa* in TQI. The latter, of course had no N-H stretching absorption at 3060cm$^{-1}$. Treat - ment of TQI with two equivalents of a reducing agent produces a blue solid which when treated with hydrogen chloride turns green with the concomitant changes observed with polyaniline including a decrease in res - istivity to $\sim 10^{-3}\Omega cm^{-1}$, a value well within that observed for dehydrated PANI.

## CONCLUSIONS

With a new preparative procedure developed in our laboratories we have been able to produce several new poly-p-phenyleneamines and an ideal model of polyaniline, a fully characterized octamer.

Surprisingly, preliminary results indicate that two of the new polymers, diaminofluorene and benzidine poly-para-phenylenediamine, could be Brönsted base doped as determined spectroscopically. This result would complete the cycle of pH-dependent changes of physical properties of polyaniline.

Our results with the phenyl-capped octamer of aniline indicate that polyaniline is just a slightly higher molecular version of this monodisperse oligomer insofar as spectroscopic structure determination  and pH-dependent transformations are concerned.

## ACKNOWLEDGMENTS

I am indebted to F.-L. Lu, M. Nowak, R. O. Angus, Jr. and P. Allemand for the research described above and to the Office of Naval Research as well as the Naval Research Laboratories through DARPA for generous support.  Thanks also go to Dr. Hugh Webb for mass spectroscopy.

# REFERENCES

1.   J. E. Frommer, Acc. Chem. Res. **19**, 2 (1986).
2.   J.-C. Chiang and A. G. MacDiarmid, Syn. Mtls. **13**, 193 (1986).
3.   D. Vachon, R. O. Angus Jr., F. L. Lu, M. Nowak, Z. X. Liu, H. Schaffer, F. Wudl and A. J. Heeger, Proceedings of the International conference on Synthetic Metals, Kyoto, 1986, Syn. Mtls. in press.
4.   R. Wilstätter and W. Moore, Chem. Ber. **40**, 2665 (1907) and references within. A. G. Green and A. E. Woodhead, J. Chem. Soc. **97**, 2388 (1910).
5.   G. E. Wnek, Polym. Prepr. **27**, 277 (1986).
6.   J. Honzl and M. Tlustakova, J. Polymer Sci. **C21**, 451(1968).
7.   J. A. Moore and J. E. Kochanowski, Macromolecules, **8**, 121 (1975).
8.   R. O. Angus, Jr., Unpublished, 1986.

# New Electronically Conducting Polymers: Effects of Molecular Structure on Intrinsic Electronic Properties

Samson A. Jenekhe
Honeywell Inc., Physical Sciences Center
Bloomington, Minnesota 55420 (U.S.A.)

## ABSTRACT

Of prime importance and interest in the continuing search for new electronically conducting polymers is the achievement of small or vanishing values of the semiconductor band gap ($E_g$) which governs the intrinsic electronic properties of materials. This paper will report on experimental studies of the effects of molecular structure on the intrinsic electronic properties of polymers. First, the structure and band gap of hetero-aromatic semiconducting polymers, including polythiophenes and polypyrroles, are discussed. It is suggested that planarity of polymer backbone is essential to small band gaps. From the results on "model compounds" for poly(2,5-thiophenediyl) it is suggested that this polymer has a non-planar backbone and an S-*cis* chain structure which largely accounts for about half of its 2.2eV band gap. Novel polybithiophenes[1] and polyterthiophenes[2] which have band gaps as small as 1.20 eV are described. Second, the synthesis and intrinsic electronic properties of a novel class of conjugated polymers containing alternating aromatic and quinonoid sequences is described[3]. These polymers exhibit band gaps as small as 0.75 eV, the smallest known value of band gap for organic polymers[4]. The quinonoid character of the polymer chains which is related to molecular parameters amenable to synthetic manipulation only partially explains why the band gap of this class of polymers is generally small. It is suggested that quantum well and superlattice effects may exist in one-dimensional conjugated polymer chains containing "mixed polymer repeating units" having different band gaps similar to inorganic semiconductor superlattice and quantum well heterostructures[5-7].

1. S.A. Jenekhe, *Macromolecules*, submitted.
2. S.A. Jenekhe, *Macromolecules*, submitted.
3. S.A. Jenekhe, *Macromolecules*, Vol. 19, in press (1986).
4. S.A. Jenekhe, *Nature 322*, 345 (1986).
5. S.A. Jenekhe, unpublished results.
6. L. Esaki and R. Tsu, *IBM J. Res. Dev. 14*, 61 (1970).
7. G.H. Dohler, "Tailored Semiconductors: Compositional and Doping Superlattices", in: T. Tsakalakos (Ed.), *Modulated Structure Materials*, Martinus Nijhoff, Dordecht, 1984; pp. 509-535.

*L. Alcácer (ed.), Conducting Polymers, 149.*
© *1987 by D. Reidel Publishing Company.*

# HIGH CONDUCTIVITY IN AN AMORPHOUS CROSSLINKED
# SILOXANE POLYMER ELECTROLYTE

R. Spindler and D. F. Shriver
Northwestern University
Department of Chemistry
Evanston, IL   60201

ABSTRACT.   A new polymer electrolyte is described which was prepared
by the reaction of poly(methylhydrosiloxane), PMHS, with poly(ethylene
glycol), PEG, and poly(ethylene glycol methyl ether), MePEG.   The
glass transition temperature of the polymer host is very low (207 K)
and the polymer forms complexes with the Li, Na, and K salts of
trifluoromethanesulfonate, all of which exhibit high ionic conductivities of up to 1 x $10^{-4}$ $\Omega^{-1}cm^{-1}$ at 40°C.   $^{29}$Si NMR, AC complex
impedance studies, DSC, and X-ray powder diffraction have been used to
characterize the electrolyte.   The temperature dependance of the conductivity has been modeled by the Vogel-Tammann-Fulcher equation.

## 1.  INTRODUCTION

An understanding of the physical properties of polymer electrolytes
can be gained by the investigation and synthesis of new host
materials.   Investigations up to now have suggested that the
successful electrolyte must contain polar groups to solvate salts,
have a low glass transition temperature, and in order to have the
highest conductivity, the electrolyte and its salt complexes should be
amorphous.   Recently a new class of polymeric hosts have been described which satisfy these requirements.[1]   A compliant phosphazene
backbone with flexible pendent polyether sidechains was employed.
This electrolyte displays high room temperature conductivity which in
most cases is 2-3 orders of magnitude greater than similar PEO complexes.   The success of the phosphazene-based electrolyte led us to
consider other polymer electrolytes with inorganic backbones.
Siloxane-based electrolytes appear to be an excellent choice due to
the low glass transition temperature of many siloxane polymers.[2]   The
siloxane-based electrolyte described here consists of a methylsiloxane
backbone with pendent polyether sidechains and crosslinks.   Other
siloxane electrolytes which have been studied include a copolymer
liquid electrolyte comprised of dimethylsiloxane and ethylene oxide
repeat units,[3] and a urethane network of poly(dimethylsiloxane grafted
ethylene oxide).[4]   Initial reports have also appeared on siloxane-

151

L. Alcácer (ed.), Conducting Polymers, 151–160.
© 1987 by D. Reidel Publishing Company.

based electrolytes related to those studied here.[5]

## 2. EXPERIMENTAL

MePEG (Aldrich, avg. MW = 350) and PEG (Aldrich, avg. MW = 300) were
dried under vacuum at 60°C for two days, no evidence of $H_2O$ was noted
by IR spectroscopy after drying.  PMHS (Petrarch, MW = 4,500-5,000)
and the catalyst Zn octoate (Petrarch, 50 wt.% PDMS) were used as
received.  Siloxane(30) was prepared by reacting a stoichiometric
amount of PEG, MePEG, PMHS and ca. 50 mg of Zn octoate/PDMS in xylene
at 130°C for ca. four hours.  Solvent was removed under vacuum and
while removing the solvent, the temperature was slowly increased to
130°C.  After ca. 2 hours, a colorless solid had formed.  We call this
polymer siloxane(30). The solid was washed for 1-4 days with $CH_2Cl_2$ in
a soxhlet extractor to remove unreacted polyethers and the catalyst.
Siloxane(30) was dried under vacuum (ca. $2 \times 10^{-5}$ torr) for > 48
hours, and stored in a dry, inert atmosphere.
 Salt complexes were prepared by weighing a stoichiometric quan-
tity of dried salt and siloxane(30) in a dry, inert atmosphere.  $CH_3CN$
(distilled from $CaH_2$) was added, and the solvent swollen polymer was
kept in contact with the salt solution for one day.  Upon removal of
the solvent, a homogeneous solid was formed and analytical data for
sulfur agrees with that calculated for polymer salt ratio employed in
the preparation found (calculated): siloxane(30) $LiSO_3CF_3$ 5% 0.99
(1.03), siloxane(30) $LiSO_3CF_3$ 10% 2.05 (2.06), siloxane(30) $LiSO_3CF_3$
15% 3.16 (3.08), siloxane(30) $LiSO_3CF_3$ 20% 3.73 (4.11)).  The complexes were dried at ca. 60°C under vacuum for 2 days and then stored
in a dry, inert atmosphere.  All concentrations of salt are reported
as weight percentages.
 $^{29}Si$ spectra were obtained on a Varian XL-400 400 MHZ instrument.
$Cr(acac)_3$, ca. 10 mM, was used as a spin relaxant and $C_6D_6$ was employed for the deuterium lock.
 DSC traces were obtained on a Perkin Elmer DSC-2 equipped with
liquid nitrogen cooling.  $T_g$'s were taken at the midpoint of the
inflection while cold crystallization exotherms and endothermic
melting transitions were measured at the peak of the transition.  All
transitions were obtained at three scanning rates and the reported
transitions were found by extrapolating to a 0°/min. heating rate.
X-ray powder diffraction traces were obtained on a Rigaku automated
powder diffractometer using CuKα radiation.
 Samples for AC impedance analysis were pressed into pellets
inside air-tight conductivity cells.  Pt or Li (ribbon, Foote Mineral
Co.) electrodes were used in these measurements.  The frequency dependent impedance of the samples was measured on a HP 4192A impedance
analyzer (13 MHz-5 Hz) or a Solartron 1250 Frequency Response
Analyzer/1286 potentiostat system (65.5 KHz-$10^{-5}$ Hz).

## 3. RESULTS AND DISCUSSION

Scheme 1 shows the reaction conditions used in the preparation of
siloxane(30).  Due to the low molecular weight of commercially avail-

able PMHS, it was necessary to crosslink the polymer with PEG to form
a solid. A crosslinking density of 30% was found to give a solid with
good mechanical properties and high ionic conductivity. The intrinsic
conductivity of siloxane(30), without added salt, is 5 x $10^{-8}$ $\Omega^{-1}cm^{-1}$
at 80°C, more than three orders of magnitude lower than the salt
complexes of siloxane(30).

### Scheme I

$$PMHS + (x/2)\ PEG + (1-x)\ MePEG \xrightarrow[\substack{Zn(octoate)_2 \\ 130°C,\ 4\ hrs.}]{xylene,\ Ar} \xrightarrow[\substack{2)\ vacuum \\ 130°C}]{1)-solvent} siloxane(30)$$

$$x = 0.30$$

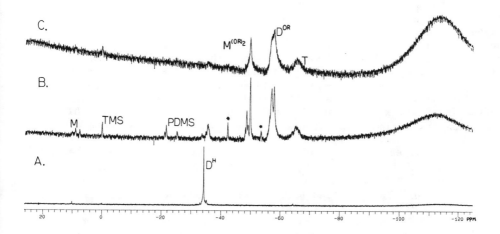

Figure 1. $^{29}$Si NMR spectra of a) PMHS, b) siloxane(30) reaction mixture after heating for four hours at 130°C; asterisks indicate
unassigned impurities, c) solvent swollen siloxane(30). The broad
feature around -110 ppm is due to the glass of the NMR tube. Standard
siloxane notation used in this paper is presented at the bottom of the
figure.

$^{29}$Si NMR is known to be an excellent probe for the molecular structure of siloxane polymers,[6] and in the present research, it has proved to be especially important. The standard siloxane nomenclature used in this paper is presented at the bottom of Figure 1. M groups are chain terminating units, D groups are chain propagating groups, and T are crosslinking moieties. The $^{29}$Si NMR spectrum of PMHS is shown in Figure 1a, one resonance is observed at - 34.1 ppm which correlates to a $D^H$ group. The reaction of PMHS with PEG and MePEG was monitored after four hours, a complex NMR spectrum was observed (Figure 1b). The resonance at -57 and -58 ppm are due to the desired product, $D^{OR}$. The peaks at -48.7 and -49.9 ppm, $M^{(OR)}_2$, and also -65 ppm, T, suggest that a significant number of $D^H$ groups are undergoing secondary reactions. These side reactions were found not to be significantly supressed by lowering the reaction temperature, changing the catalyst ($K_2CO_3$, $ZnCl_2$, or $SnCl_2$), or by switching solvents to THF and lowering the catalyst concentration. This last set of conditions was the same used by Smid and coworkers.[5a] It is known that metal hydroxides and metal alkoxides, in the presence of alcohols, cleave Si-O-Si bonds to form alkoxy substituted siloxanes and $H_2O$.[7] If such a reaction is taking place during the preparation of siloxane(30) it would cause the formation of the T and $M^{(OR)}_2$ moieties. The crosslinked solid was also investigated by $^{29}$Si NMR. The NMR spectrum of the solvent swollen polymer (Figure 1c) shows that the $D^{OR}$, $M^{(OR)}_2$, and T groups are still present, and furthermore that the extraction with $CH_2Cl_2$ was efficient in removing PDMS and the unassigned impurities.

The thermal properties of the starting materials used in the synthesis of siloxane(30) and its salt complexes are shown in Table I. Low glass transition temperatures, $T_g$, have been strongly correlated to high ionic conductivity in polymer electrolytes, the siloxane(30) electrolytes used in this study corroborate this concept. The change in $T_g$ upon addition of salt is very slight, even adding 20% of $LiSO_3CF_3$ only increases the glass transition temperature by 23 K. Addition of 5.5% $NaSO_3CF_3$ and 6.0% $KSO_3CF_3$ causes no appreciable change in $T_g$ within our experimental error. Siloxane (30), siloxane(30) $LiSO_3CF_3$ 5%, and siloxane(30) $NaSO_3CF_3$ 5.5% all exhibit a cold crystallization after the glass transition which is followed by the melting of the crystalline phase. When the siloxane(30) and siloxane(30) $LiSO_3CF_3$ 5% samples are quenched at 10°/min, the cold crystallization exotherm is no longer observed. Addition of more $LiSO_3CF_3$ tends to avert the formation of the crystalline phase. In the case of siloxane(30) $NaSO_3CF_3$ 5.5%, the cold crystallization exotherm is broader and weaker than for the $LiSO_3CF_3$ salt, and also the cold crystalization is suppressed when the sample is cooled at 5°/min. As Na$^+$ is replaced by the heavier K$^+$ (the siloxane(30) $LiSO_3CF_3$ 5%, siloxane(30) $NaSO_3CF_3$ 5.5%, and siloxane(30) $KSO_3CF_3$ 6% complexes all contain the same number moles of salt per gram of polymer), the cold crystallization exotherm and the melting endotherm are no longer observable. The cold crystallization and melting transitions observed for some of these systems is reminiscent of the behavior of PEG and MePEG (Table I), this suggests that the crystallization is associated with the polyether side groups.[8] X-ray powder

diffraction verifies that the siloxane(30) polymer and salt complexes
are amorphous at ambient temperatures.

### TABLE I  DSC and Conductivity (40°C) DATA

| Polymer | $T_g(K)$ | $T_{cc}(K)$ | $T_m(K)$ | $\sigma_{40}$* |
|---|---|---|---|---|
| PMHS | 133 | -- | -- | -- |
| PEG | 194 | 213 | 255 | -- |
| MePEG | 178 | 205,221 | 246,266 | -- |
| siloxane(30) | 207 | 218 | 265 | $< 10^{-9}$ |
| siloxane(30) $LiSO_3CF_3$ 5% | 210 | 278 | 267 | $3.1 \times 10^{-5}$ |
| siloxane(30) $LiSO_3CF_3$ 10% | 213 | -- | -- | $5.3 \times 10^{-5}$ |
| siloxane(30) $LiSO_3CF_3$ 15% | 224 | -- | -- | $7.3 \times 10^{-5}$ |
| siloxane(30) $LiSO_3CF_3$ 20% | 230 | -- | -- | $5.2 \times 10^{-5}$ |
| siloxane(30) $NaSO_3CF_3$ 5.5% | 208 | 234 | 257 | $4.9 \times 10^{-5}$ |
| siloxane(30) $KSO_3CF_3$ 6.0% | 207 | -- | -- | $1.0 \times 10^{-4}$ |

*  Conductivity at 40°C in units of $\Omega^{-1}cm^{-1}$.

The electrical properties of the siloxane(30) complexes were
probed by complex impedance measurements from 5 MHz to 10 mHz
employing ion-blocking Pt electrodes or cation reversible Li electrodes. When Pt electrodes were used, the complex impedance spectra
showed the expected semicircle and low frequency spur for the
siloxane(30) electrolytes (Figure 2a).[9] The low frequency spur is
replaced by a small semicircle when Li electrodes are used (Figure
2b), this arc is attributed to an ionic charge transfer process
between the polymer electrolyte and the lithium electrode.[9,10] The
charge transfer arc is not observable above 60°C. The siloxane(30)
electrolytes were found to be stable to Li metal at temperatures up to
100°C and no significant changes in the complex impedance spectra were
seen upon temperature cycling. Our studies of the concentration

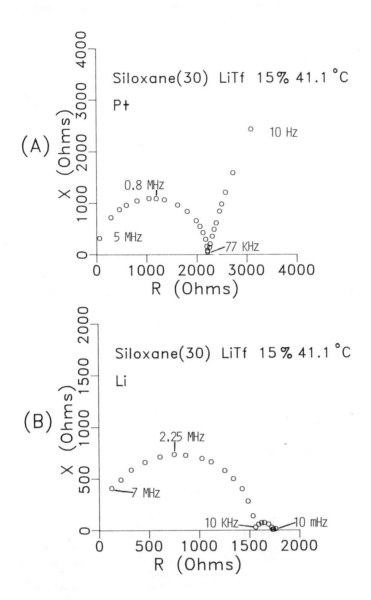

Figure 2. Complex impedance spectra of siloxane(30) LiSO$_3$CF$_3$ 15% with
a) Pt and b) Li electrodes.

dependence of the conductivity, for the siloxane(30)/LiSO$_3$CF$_3$ system,
indicate that the highest conductivity is found for siloxane(30)
LiSO$_3$CF$_3$ 15% electrolyte (Table I), similar concentration dependence
has been noted for other amorphous electrolytes.[1,4]  Changing the

alkali metal cation from $Li^+$ to $Na^+$ and to $K^+$ had a noticeable effect
on the conductivity of the electrolytes (Table I).  The highest
conductivity is observed with the $KSO_3CF_3$ doped electrolyte (1 x $10^{-4}$
$\Omega^{-1}cm^{-1}$ at 40°C) and the conductivities decrease as we replace $K^+$ with
the lower MW alkali metal cations.  Similar trends have been observed
for polymer electrolytes[11] and also for nonaqueous electrolytes, where

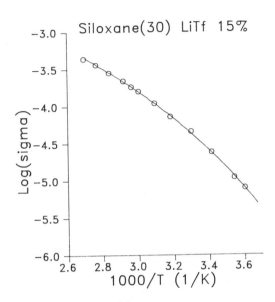

Figure 3. Temperature dependence of the ionic conductivity of
siloxane(30) LiSO $_3CF_3$ 15%.  The line drawn through the data points
was calculated using the VTF equation.

the mobility of the cation tends to increase with increased ionic
radius of the cation.[12]  Figure 3 shows a plot of log ($\sigma$) vs 1000/T
for siloxane(30) $LiSO_3CF_3$, which displays a distinct curvature.  This
temperature dependance has been documented for amorphous polymer elec
electrolytes[1,11,13] and for aqueous salt solutions.[15]  The Vogel-
Tammann-Fulcher (VTF)[14] relationship, eq. 1, has been employed to
successfully model the temperature dependence of conductivity for many
amorphous polymer electrolytes.[4,13a,13b,16]  The conductivity data was
fit to the VTF relationship by a least squares method with successive
iterations of $T_0$ until

$$\sigma \quad = \quad AT^{-1/2} \quad \exp \quad \frac{-E_a}{T-T_0} \tag{1}$$

a correlation coefficient of $\geq$ 0.999 was obtained.  The coefficients
obtained are presented in Table 2.

## VTF Parameters

(A)

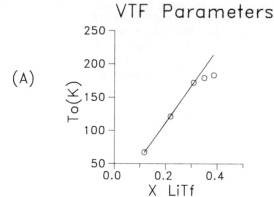

(B)

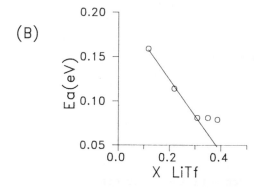

Figure 4. Plots of $T_0$ and $E_a$ versus mole fraction of $LiSO_3CF_3$ contained in the electrolyte. The parameters $T_0$ and $E_a$ were derived by fitting the VTF equation to the experimental conductivity data.

### TABLE II   VTF Parameters

| sample | $T_0(K)$ | $T_g-T_0(K)$ | $E_a(eV)$ |
|---|---|---|---|
| siloxane(30) $LiSO_3CF_3$ 5% | 67 | 173 | 0.159 |
| siloxane(30) $LiSO_3CF_3$ 10% | 121 | 92 | 0.114 |
| siloxane(30) $LiSO_3CF_3$ 15% | 172 | 52 | 0.081 |
| siloxane(30) $LiSO_3CF_3$ 20% | 183 | 47 | 0.079 |
| siloxane(30) $NaSO_3CF_3$ 5.5% | 126 | 82 | 0.113 |
| siloxane(30) $KSO_3CF_3$ 6.0% | 162 | 45 | 0.083 |

It is evident that the $T_0$ value for the siloxane(30) $LiSO_3CF_3$ 5%, $LiSO_3CF_3$ 10%, and $NaSO_3CF_3$ 5.5% do not exhibit the expected behavior of $T_g-T_0$ = 50K. The apparent activation energies ($E_a$) decrease upon increasing the concentration of salt while $T_0$ increases upon increasing salt concentration. Figure 4 shows plots of $T_0$ vs $X_{LiTf}$ (mole fraction) and $E_a$ vs. $X_{LiTf}$. Even though the number of data points is not sufficient to get a clear picture, it appears that both $T_0$ and $E_a$ have a linear dependance upon $X_{LiTf}$ over the composition range from $LiSO_3CF_3$ 5% to $LiSO_3CF_3$ 15%. A similar linear dependance of $T_0$ and $E_a$ has been observed for concentrated aqueous $Ca(NO_3)_2$[17] and formamide $NaI$[18] solutions. These results do point out the utility of comparing polymer electrolytes to concentrated aqueous and nonaqueous salt solutions.

## 4. CONCLUSIONS

The ionic conductivity of the siloxane-based electrolyte described here are as high as any known polymer electrolyte system, and in fact the conductivity is close to that seen in liquid polyether salt solutions. Further increases in conductivity may require a radical change in the nature of the polymer or the salt.

## 5. ACKNOWLEDGEMENTS

This research was sponsored by the NSF MRL program Grant No. DMR 8216972 through the Northwestern University Materials Research Center. L. C. Hardy is acknowledged for initial investigations in this area and for helpful discussions.

## 6. REFERENCES

1.    Blonsky, P. M.; Shriver, D. F.; Austin, P.; Allcock, H. R. J. Am. Chem. Soc. 1984, 106, 6454.

2.    Noll, W. "Chemistry and Technology of Silicones", Academic Press: New York, 1968; pp. 326-327.

3.    Nagaoka, K.; Naruse, H.; Shinohara, D.; Watanabe, M. J. Polym. Sci. Polym. Lett. Ed. 1984, 22, 659-662.

4.    Bouridah, A.; Dallard, F.; Deroo, D.; Cheradame, H.; LeNest, J. F. Solid State Ionics 1985, 15, 233-240.

5.    (a) Fish, D.; Khan, D. M.; Smid, J. Polym. Prepr. 1986, 27, 325-326. (b) Hall, P. G.; Davies, G. R.; McIntyre, J. E.; Ward, D. M.; Bannister, D. J.; LeBrocq, K. M. F. Polym. Commun. 1986, 27, 98-100.

6.    (a) Harris, R. K.; Kennedy, J. D.; McFarlene, W.; In "NMR and the Periodic Table"; Harris, R. K.; Mann, B. E. Eds.; Academic Press: New York, 1978; pp. 309-377. (b) Harris, R. K.; Kimber, B. J. Appl.

Spectrosc. Rev. **1975**, <u>10</u>, 117–137.   (c)   Beshah, K.; Mark, J. E.; Ackerman, J. L.; Himstedt, A.   J. Polym. Sci. Polym. Phys. **1986**, <u>24</u>, 1207–1225.

7.   (a) Voronokov, M. G.; Shabarova, Z. D.   J. Gen. Chem. USSR **1959**, <u>29</u>, 1501–1507.   (b)   Voronokov, N. G.; Mileshkevich, V. P.; Yuzhelevskii, Y. A.   "The Siloxane Bond"; Consultants Bureau:   New York, **1978**; pp. 149–157.

8.   Wunderlich, B.   "Macromolecular Physics" Vol. 3; Academic Press: New York, 1980; pp. 325–331.

9.   Macdonald, J. R.   J. Chem. Phys. **1974**, <u>61</u>, 3977–3996.

10.   (a) Sørensen, P. R.; Jacobsen, T.   Electrochem. Acta **1982**, <u>27</u>, 1671–1675.   (b) Hardy, L. C.; Shriver, D. F.   J. Am. Chem. Soc. **1985**, <u>107</u>, 3823–3828.   (c)   Stainer, M.; Hardy, L. C.; Whitmore, D. H.; Shriver, D. F. J. Electrochem. Soc. **1984**, <u>131</u>, 784–790.

11.   Watanabe, M.; Sanui, K.; Ogata, N.; Inoue, F.; Kobayashi, T.; Ohtaki, Z.   Polym. J. **1985**, <u>17</u>, 549–555.

12.   (a) Spiro, M.   In "Physical Methods of Chemistry" Vol. 2; Rossiter, B. W.; Hamilton, J. F., Eds.; John Wiley & Sons:   New York, 1985:   pp 685–686.   (b)   Fernandez-Prini, F.   In "Physical Chemistry of Organic Solvent Systens"; Covington, A. K.; Dickinson, T., Eds.; Plenum Press:   London, 1973; pp. 587–614.

13.   (a) Armand, M. B.; Chabagno, J. M.; Duclot, M. J.   In "Fast Ion Transport in Solids Electrodes and Electrolytes"; Elsevier North Holland:   New York, 1979; pp. 131–136.   (b)   Cheradame, H.   In "IUPAC Macromolecules"; Benoit, H.; Tempp, P., Eds.;   Pergamon Press: Oxford, 1982; pp. 251–264.   (c)   Spindler, R.; Shriver, D. F. Macromolecules **1986**, <u>19</u>, 347–350.

14.   (a) Vogel, H.   Phys. Z **1921**, <u>22</u>, 645.   (b) Tammann, G.; Hesse, W. Z. Anorg. Allg. Chem.   **1926**, <u>156</u>, 245–257.   (c)   Fulcher, G. S.   J. Am. Ceram. Soc. **1925**, <u>8</u>, 339–355.

15.   Angel, C. A.   J. Phys. Chem. **1966**, <u>70</u>, 3988–3998.

16.   (a)   Harris, C. S.; Shriver, D. F.; Ratner, M. A. Macromolecules, **1986**, <u>19</u>, 987–989.   (b)   Fontanella, J. J.; Wintersgill, M. C.; Smith, M. K.; Semancik, J.   J. Appl. Phys.   In press.

17.   Angel, C. A.; Bressel, R. D.   J. Phys. Chem. **1972**, <u>22</u>, 3244–3253.

18.   Bruno, P.; Gatti, C.; Monica, M. D.   Electrochimica Acta **1975**, <u>20</u>, 533–538.

# USE OF ELECTRONICALLY CONDUCTING POLYMERS AS CATALYTIC ELECTRODES IN AQUEOUS AND INORGANIC ELECTROLYTES

Robert J. Mammone
US Army Electronics Technology and Devices Laboratory
Power Sources Division
Fort Monmouth, NJ  07703-5000

ABSTRACT.  Polyacetylene can be oxidized ("p-doped") to the metallic regime by $O_2$ or $H_2O_2$ in the presence of an aqueous solution of a non-oxidizing acid such as $HBF_4$ or by an aqueous solution of an oxidizing acid such as $HClO_4$.  The $[(CH)^{+y}(A)^-y]_x(A^- = ClO_4^-$ or $BF_4^-)$ formed above may be used as a cathode by connecting it to a lead anode in the electrolyte whereupon it is reduced to $(CH)_x$.  If the oxidizing agent is present during this process, the $(CH)_x$ is continuously electrochemically reduced by the lead as rapidly as it is oxidized chemically by the oxidizing agent.  Polyacetylene is therefore acting as a catalytic electrode.

Electrochemical characteristics of polythiophene and poly-3-methylthiophene have been investigated in liquid $Li(SO_2)_3AlCl_4$ electrolyte using cyclic voltammetry.  These polymers were electropolymerized on smooth vitreous carbon rods and cycled between 2.0 and 4.0 volts relative to lithium.  Although both polymers showed a 2.9 volt reduction wave (due to $SO_2$ reduction), only the poly-3-methylthiophene showed significant oxidation at 3.9 volts.  Use of poly-3-methylthiophene as a catalytic electrode is therefore of interest to rechargeable $Li/SO_2$ cell technology.

## 1.  INTRODUCTION

Until recently, the practical application of polyacetylene has been restricted to uses where oxygen and water could be rigorously excluded since both neutral $(CH)_x$ and its oxidized or reduced forms were believed to react with these substances rapidly and irreversibly with complete loss of electrical conductivity (1).  However, it was recently demonstrated that neutral $(CH)_x$ may be electrochemically oxidized ("p-doped") and then reduced in strong (7.4 M) aqueous $HBF_4$ solution (2-3) or in a saturated (5.6 M) aqueous $Pb(ClO_4)_2$ solution (4).  In addition, a more detailed understanding of the oxidation ("p-doping") process by which the conductivity of cis- or trans-$(CH)_x$ can be increased through the semiconducting to the metallic regime has indicated that gaseous oxygen will "dope" $(CH)_x$, under appropriate experimental conditions, to the metallic regime (2,3,5).  In the

*L. Alcácer (ed.), Conducting Polymers, 161–172.*

absence of acid, oxygen reacts irreversibly with the semiconducting $(CH)_x$ to produce an insulating material of unknown, complex composition containing C=O and other O-containing groups (1). The $[(CH)^{+y}(A)^-y]_x$ ($A^- = BF_4^-$) formed by chemical oxidation of neutral $(CH)_x$ with oxygen may be used as a cathode in an $HBF_4$ electrolyte. If this polymer is now connected to a lead anode through an external load, the polymer is reduced to $(CH)_x$ and an electric current flows through the external circuit. The electric current will continue to flow as long as oxygen is present to chemically oxidize the $(CH)_x$. The $(CH)_x$, therefore, acts as a catalytic electrode for the reduction of gaseous oxygen. Further studies (6) have been carried out on gaseous oxygen and this concept has been applied to liquid oxidants including $H_2O_2$ and $HClO_4$. The present work summarizes the electrocatalytic properties of polyacetylene with gaseous $O_2$, $H_2O_2$ and $HClO_4$ in aqueous electrolytes.

The concept of employing conducting polymers as catalytic electrodes for reduction of liquid oxidants can be extended to include inorganic electrolytes as well. Reports in the literature have been scarce and deal primarily with use of conducting polymers in primary oxyhalide systems (7-8). Recently, considerable amount of in-house research has gone into studying applications of polymers as catalytic electrodes in rechargeable Li/SO_2 cells employing $Li(SO_2)_3AlCl_4$ as the electrolyte. At room temperature and atmospheric pressure, this electrolyte has a measured conductivity of 0.13 S/cm (9). Reduction and oxidation processes occuring at smooth carbon electrodes in $Li(SO_2)_3AlCl_4$ were studied by Mammone and Binder (9). They hypothesized that since no clear oxidation wave for the insoluble reduced product is seen at smooth carbon rods in $Li(SO_2)_3AlCl_4$, free molecular chlorine formed from $AlCl_4^-$ oxidation during charge chemically re-oxidizes the product. Charging potentials of such cells is typically at 3.9 - 4.0 volts (10). While oxidizing agents are a useful and necessary intermediate product of cell charging, use of chlorine as an intermediate nevertheless poses battery engineering and safety problems. Nevertheless, use of this electrolyte in lithium/sulfur dioxide rechargeable cells in which carbon black has been used as the cathode substrate, has recently been actively studied (9-11).

Since this electrolyte possesses the highest known conductivity of any ambient temperature non-aqueous electrolyte, it would be of interest to investigate the potential applications of conducting polymers in this system. One potential application would be to develop a rechargeable SO_2 cell employing a conducting polymer as the cathode substrate. Use of a selected polymer which oxidizes at potentials between 3.0 and 4.0 volts would be of interest since it would preclude formation of molecular chlorine during charge. The oxidized polymer would then oxidize the cathode discharge product and thereby become reduced itself. In this process, the cathode discharge product would still become oxidized during charge but charging potentials would be insufficient to generate molecular chlorine. Another potential application would be to use a selected polymer as the cathode-active material in place of the sulfur dioxide in this highly conducting $Li(SO_2)_3AlCl_4$ electrolyte. Since thin films of many polymers can be readily prepared, a cell constructed in this electrolyte could potentially have extremely high

power densities which could be of interest to the military for certain
applications.  However, the polymer will be required to undergo oxidation and reduction between 3.0 and 4.0 volts.

A conducting polymer chosen for these potential applications was
polythiophene and its 3-methyl derivative.  These polymers represent a
class of tractable, conducting materials which exhibit facile and
reversible electrochemistry.  When in the oxidized form, these polymers
become metallic ( $\sigma \sim 10^0 - 10^2$ S/cm).  Thin films of polythiophene and
its derivatives supported on various substrates can be made in situ
from the appropriate monomers.  Polythiophenes are additionally attractive from the standpoint that their electrochemical and physical
properties are readily altered via substituent groups (12).    The
present work investigates the electrochemical properties of polythiophene and its 3-methyl derivative in liquid $Li(SO_2)_3AlCl_4$.


2.  EXPERIMENTAL

Polyacetylene film was synthesized as described previously (13).
Flouroboric acid, 48% (Aldrich) and perchloric acid, 70% (Fischer) were
used without further purification or dilution.  Lead metal (Alfa Ven-
tron) was scraped with a knife immediately before use.  Glass filter
paper (Reeve Angle 934AH) was employed as the separator in the studies
of the electrocatalytic properties of $(CH)_x$ in $HClO_4$.

For the $O_2$- and $H_2O_2$-electrocatalytic studies, cis-rich $(CH)_x$ was
first electrochemically oxidized to $[(CH)^{+0.06}(BF_4)^-_{0.06}]_x$ in a 0.5 M
solution of $(n-C_4H_9)_4NBF_4$ in methylene chloride as described previously
(5).  For both investigations, a strip of $[(CH)^{+0.06}(BF_4)^-_{0.06}]_x$ (1.0 x
1.0 x 0.01 cm), to the top of which a platinum wire had been attached,
was placed in one leg of a polyethylene H-cell containing a solution of
48% $HBF_4$.  The platinum wire attached to the film and the area of the
film touching the platinum was covered with molten paraffin wax.   A
lead counter electrode (1.0 x 4.0 x 0.013 cm) was placed in the solution in the other leg of the H-cell at a distance of $\sim 5$ cm from the
$[(CH)^{+0.06}(BF_4)^-_{0.06}]_x$ electrode.   Electrochemical reduction of the
$[(CH)^{+0.06}(BF_4)^-_{0.06}]_x$ was accomplished by having it serve as the
cathode by holding it at a constant potential of 0.01 volts with respect to the lead counter electrode for 1 hour.   Current versus time
curves were recorded continuously for the 1 hour electrochemical reduction of the $[(CH)^{+0.06}(BF_4)^-_{0.06}]_x$ and for all subsequent studies.

For the $HClO_4$-electrocatalytic studies, the $(CH)_x$ electrode was
made from cis-rich $(CH)_x$ film (1.0 x 1.0 x 0.01 cm) which was mechanically pressed onto 52 mesh platinum gauze.  The mesh was then folded in
half with the $(CH)_x$ being inside.  The cell was constructed by placing
two pieces of glass filter paper between the $(CH)_x$ electrode and the
lead electrode (1.0 x 4.0 x 0.013 cm), squeezing the assembly into 3 x
4 x 30 mm rectangular glass tubing (Vitro Dynamics, Inc.), adding the
electrolyte and sealing the cell with molten paraffin wax across the
protruding electrode leads.  The cell was allowed to stand for 18 hours
before the experiments were commenced.

Polythiophene and poly-3-methylthiophene were prepared by electro-

chemical    polymerization    of    the    appropriate    monomers    in    a
three-electrode, single compartment apparatus.  The films were grown on
smooth 0.073 cm$^2$ vitreous carbon rods produced at 1000$^\circ$C (Atomergic)
with a platinum counter-electrode and an Ag/Ag$^+$ (0.1 M AgNO$_3$/CH$_3$CN)
reference electrode.  The solutions used for the preparation contained
0.1 M of the appropriate monomer with 0.1 M (n-C$_4$H$_9$)$_4$NBF$_4$ in acetonitrile.  The acetonitrile (Burdick and Jackson) was dried over 4A
molecular sieves prior to use.

        The Li(SO$_2$)$_3$AlCl$_4$ electrolyte was prepared by finely grinding
white LiAlCL$_4$ (Anderson Physics) in an argon filled dry box and allowing SO$_2$ gas to pass over the solid until all the LiAlCl$_4$ has reacted to
form a lightly straw colored low vapor pressure liquid ( $\sim$ 15 min.).
Neutrality of the resulting liquid complex was assured by adding excess
LiCl and allowing the liquid to equilibrate overnight at room temperature.

        A three-electrode design using both a lithium reference and
counter-electrode was used for cyclic voltammetry experiments.  Electrolyte volume was 12.5 ml.  Lower and upper voltage scan limits were
from 2.0 to 4.0 volts relative to lithium with initial scan directions
being cathodic of respective open circuit potentials.  All data were
obtained at room temperature and 100 mV/sec scan rates and referenced
to a small lithium strip also immersed in the electrolyte.  The lithium
counter and reference electrodes remained shiny in all electrolytes
studied and the potential of the SO$_2$ reduction peak at 2.9 volts was
essentially unchanged in all the solutions studied.  This implies that
at least over a one day time scale, no gross lithium corrosion was
taking place.

        Constant potential or constant current oxidation and reduction
(for electrocatalytic studies) were carried out utilizing a Princeton
Applied Research (PAR) Model 363 potentiostat-galvanostat in conjunc-
tion with an Electro-Synthesis Co. (ECS) Model 630 digital coulometer.
Current versus time or voltage versus time curves were recorded with a
Houston    Instruments    series    D500    strip-chart    recorder.    Cyclic
voltammetry studies were carried out utilizing a Princeton Applied
Research (PAR) Model 173 potentiostat-galvanostat in conjunction with a
Model 175 universal programmer.  Current-voltage curves were recorded
with a Hewlett-Packard Model 7047A x-y recorder.


3.  RESULTS AND DISCUSSIONS

3.1  Use of Polyacetylene as a Catalytic Electrode in Aqueous Electrolytes

3.1.1.  <u>Oxygen and Hydrogen Peroxide</u>.  The effect of gaseous oxygen and
hydrogen peroxide on the current produced on a 1 cm$^2$ (CH)$_x$ electrode is
shown in Figure 1 and 2, respectively.  The initial current, given in
the decreasing left hand curves is the initial electrochemical reduction of the [(CH)$^{+0.06}$(BF$_4$)$^-_{0.06}$]$_x$.  When the film had been almost
completely reduced to (CH)$_x$, the oxidant was added and the current was
again recorded.  This resulted in the continuous <u>chemical</u> oxidation and

electrochemical reduction of the partly oxidized polyacetylene electrode. In the case of gaseous oxygen, the current increased from 0.029
mA to a maximum of 0.47 mA after 90 minutes and then slowly decayed to
0.29 mA after 24 hours of continuous operation. The amount of charge
released during this time was 33.18 C which corresponded to the passage
of 1.702 electrons per CH unit in the $(CH)_x$. In the case of hydrogen
peroxide, the current increased from 0.039 mA to a maximum of 3.4 mA
after 2.5 minutes and then slowly decayed to 2.3 mA after 4 hours of
continuous operation. The amount of charge released during this time
was 38.05 C which corresponds to the passage of 2.062 electrons per CH
unit in the $(CH)_x$. The electrocatalytic effects are obvious in these
experiments since more than 1 electron was passed per CH unit.

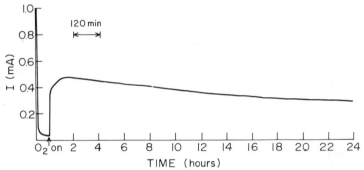

Figure 1.    Change in current produced by a $(CH)_x/O_2/48\%$ $HBF_{4(aq)}/Pb$
cell when oxygen stream is continuously bubbled over the
$(CH)_x$ electrode.

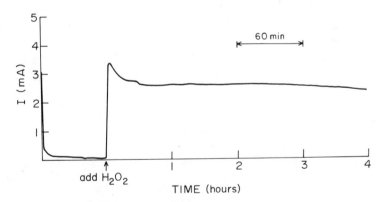

Figure 2.    Change in current produced by a $(CH)_x/O_2/48\%$ $HBF_{4(aq)}/Pb$
cell when enough hydrogen peroxide is added to the cell to
make a 0.1 M solution.

It has been show previously that gaseous oxygen or hydrogen peroxide will oxidize $(CH)_x$ to the metallic regime when dissolved in a
strong, non-oxidizing acid such as 48% $HBF_4$ (2,5). The overall reac-

tion can be expressed for oxygen by the following equation:

$$4(CH)_x + 4(xy)HBF_4 + (xy)O_2 \longrightarrow 4[(CH)^{+y}(BF_4)^-_y]_x + 2(xy)H_2O \qquad (1)$$

The oxygen oxidizes the $(CH)_x$ to $[(CH)^{+y}]_x$ $(y \sim 0.02)$ and the $(BF_4)^-$ acts as the stable counter ion. If a piece of $[(CH)^{+0.02}(BF_4)^-_{0.02}]_x$ film and a strip of lead are placed in 48% $HBF_4$ and are connected via an external wire, the lead dissolves, liberating electrons which are taken up by the $[(CH)^{+0.02}]_x$ ion resulting in the net electrochemical reduction reaction:

$$(0.01x)Pb + [(CH)^{+0.02}(BF_4)^-_{0.02}]_x \longrightarrow (CH)_x + (0.01x)Pb(BF_4)_2 \qquad (2)$$

which regenerates the $(CH)_x$. The lead, which acts as the reducing agent, is converted to $Pb(BF_4)_2$. If oxygen is constantly bubbled over the polyacetylene electrode, it is possible to continuously <u>chemically</u> oxidized the $(CH)_x$ to $[(CH)^{+0.02}(BF_4)^-_{0.02}]_x$ as rapidly <u>as it is</u> reduced <u>electrochemically</u> according to equation (2). The overall reaction which occurs can be expressed by the following equation:

$$Pb + 1/2O_2 + 2HBF_4 \xrightarrow{(CH)_x} Pb(BF_4)_2 + 2H_2O \qquad (3)$$

The $(CH)_x$ is, therefore, acting as an electrocatalytic electrode for the reduction of oxygen.

As can be seen, analogous electrocatalytic effects in aqueous solution with the concomitant generation of an electric current are observed when hydrogen peroxide is used as the oxidizing agent instead of oxygen. The overall reaction which occurred can be expressed by the following equation:

$$H_2O_2 + 2HBF_4 + Pb \xrightarrow{(CH)_x} Pb(BF_4)_2 + 2H_2O \qquad (4)$$

In this study, the current remained essentially constant during the course of the experiment.

3.1.2. <u>Perchloric Acid</u>. The effect of perchloric acid on a 1 $cm^2$ $(CH)_x$ electrode for constant current discharges of 1.0 mA and 10.0 mA is shown in Figure 3. The cells were allowed to stand for 18 hours before the constant current discharges commenced. The open circuit voltage of the cells increased to $\sim 0.9$ volts during this time, indicating that the $(CH)_x$ was being chemically oxidized by the aqueous 12 M $HClO_4$. The cells were then discharged until the potential fell to 0.2 volts. Both discharge curves showed a characteristic "plateau" in which the voltage changes only slightly with time. However, there was a sharp decrease in voltage at the end of the discharges which indicates that either the $(CH)_x$ electrode was passivated with an insoluble reduced product and/or some degradation of the $(CH)_x$ electrode had taken place. The capacity of the $(CH)_x$ electrode was 14.2 mAhr/$cm^2$ and 29.3 mAhr/$cm^2$ for the 1 mA/$cm^2$ and 10 mA/$cm^2$ discharges, respectively.

This corresponds to the passage of 2.468 and 4.064 electrons per CH unit in the $(CH)_x$. After completion of the constant current dischar- ges, a white precipitate of $PbCl_2$ (identified by x-ray powder pattern) was observed on the $(CH)_x$ electrode.

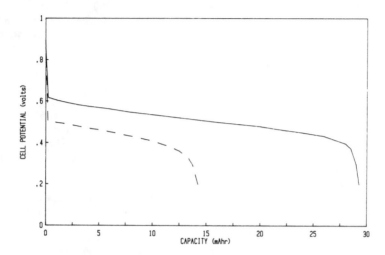

Figure 3.  Constant current discharges of a $(CH)_x/70\%$ (12 M) $HClO_{4(aq)}/Pb$ cell at 1.0 mA (solid line) and at 10 mA (dashed line).

It has been shown previously from reduction potentials of $(CH)_x$, $[(CH)^{+y}]_x$ and $HClO_4$, that $(CH)_x$ is oxidized by $HClO_4$ (2). The overall reaction can be expressed by the following equation:

$$8(CH)_x + 9(xy)HClO_4 \longrightarrow 8[(CH)^{+y}(ClO_4)^-_y]_x + (xy)HCl + 4(xy)H_2O \quad (5)$$

If a piece of cis-$(CH)_x$ film and a strip of lead are placed in 12 M $HClO_4$, the $(CH)_x$ is oxidized to $[(CH)^{+y}]_x(y \sim 0.1)$ according to equation (5) and the lead does not dissolve spontaneously to any significant extent. If, however, these two electrodes are now connected via an external wire, the lead dissolves, liberating electrons which are taken up by the $[(CH)^{+0.1}]_x$ ion resulting in the net electrochemical reduc- tion reaction:

$$(0.05x)Pb + [(CH)^{+0.1}(ClO_4)^-_{0.1}]_x \longrightarrow (CH)_x + (0.05x)Pb(ClO_4)_2 \quad (6)$$

which regenerates the $(CH)_x$. The lead, which acts as the reducing agent, is converted to $Pb(ClO_4)_2$. It should be noted that the $(CH)_x$ is reconverted back to $[(CH)^{+0.1}(ClO_4)^-_{0.1}]_x$ by the 12 M $HClO_4$ as given by equation (5). Hence, it is possible to continuously chemically oxidize the $(CH)_x$ to $[(CH)^{+0.1}(ClO_4)^-_{0.1}]_x$ as rapidly as it is reduced electro- chemically according to equation (6). The overall reaction which occurs can be expressed by the following equation:

$$16HClO_4 + 8Pb \xrightarrow{\text{(CH)}_x} PbCl_2 + 7Pb(ClO_4)_2 + 8H_2O \qquad (7)$$

The $(CH)_x$ is therefore acting as an electrocatalytic electrode for the reduction of perchloric acid.

It is clear that the operation of the $(CH)_x$ electrode in this system has a completely different behavior from that of a conducting polymer or inorganic secondary battery electrode. Unlike these systems which undergo direct reduction during cell discharge and direct oxidation during charge, the $(CH)_x$ electrode simply acts as a matrix capable of accomodating the cell reduction products. In this regard, the operation of the $(CH)_x$ electrode is very similar to a carbon black electrode utilized in liquid oxyhalide systems (e.g. $SOCl_2$ and $SO_2Cl_2$) since during reduction of the liquid oxidant, an insoluble discharge product is deposited in the pores of the cathode substrate.

## 3.2. Cyclic Voltammetric Studies of Polythiophene and Poly-3-methyl-thiophene in $Li(SO_2)_3AlCl_4$

Cyclic voltammograms at 100 mV/sec for thin electrodeposited polythiophene and poly-3-methylthiophene films on smooth carbon rods in conducting liquid $Li(SO_2)_3AlCl_4$ electrolyte at room temperature are shown in Figure 4. A voltammogram of the bare carbon rod is also shown for comparison. The dashed lines in Figure 4 are the second scans obtained without cleaning the smooth carbon rod or removal of the deposited product. Figure 4a shows cyclic voltammograms on the bare carbon rod in $Li(SO_2)_3AlCl_4$. The reduction wave at 2.9 volts is believed to be due to reduction of $SO_2$ (9). The observation that no distinct oxidation wave exists for the 2.9 volt reduction product and the fact that the carbon electrode becomes passivated implies that formation of the reduction product at 2.9 volts is essentially irreversible. Ordinarily during cell charging, the cell potential rises to 4.0 volts. This is due to molecular chlorine formed at that potential which is responsible for chemically oxidizing the cathode discharge product.

Figures 4b and 4c show cyclic voltammograms of polythiophene and poly-3-methylthiophene-coated carbon electrodes. There are striking differences between the voltammograms for the two polymer coated electrodes and the uncoated baseline. The most striking feature of these voltammograms is that the intensity of the reduction wave at 2.9 volts of the poly-3-methylthiophene-coated electrode was about 50 times as large as the principle reduction wave seen in the case of the bare smooth carbon alone. The intensity of the reduction wave at 2.9 volts of the polythiophene-coated electrode was only about 3.5 times as large as the principle reduction wave seen in the case of the bare smooth carbon rod alone. The reason for the enhanced reduction and oxidation currents in the case of these polymers, in general, can be explained by the fact that the in situ polymerized coatings are thin, porous in nature and possess high surface areas. The higher active surface area gives larger electrode capacitances which in turn produce high slopes in measured I-V curves. More surface area provides additional sites for electrochemical reaction and results in huge measured currents.

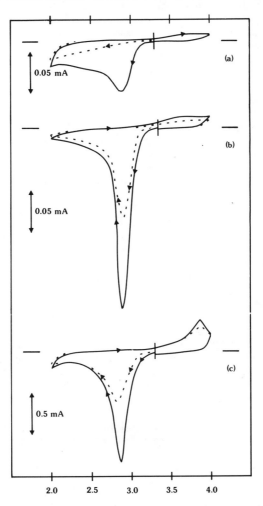

POTENTIAL RELATIVE TO LITHIUM, VOLTS

Figure 4.   First and second cyclic voltammgrams (100 mV/sec) obtained
            by scanning cathodically from the OCV in $Li(SO_2)_3AlCl_4$ on
            (a) a smooth carbon rod, (b) a smooth carbon rod coated with
            polythiophene and (c) a smooth carbon rod coated with poly-
            3-methylthiophene.

Both the polythiophene and poly-3-methylthiophene coated electrodes
exhibited reduction currents on the first scan substantially larger
than those of the second scan.  This is due to passivation of the electrode by the insoluble reduction product.  In addition, both polymers
appeared to be stable in the electrolyte.

The voltammogram for the polythiophene-coated electrode showed a principle reduction wave centered about 2.9 volts. This is primarily due to $SO_2$ reduction. However, no well defined oxidation peak for polythiophene is seen. Rather a small gradual sloping oxidation wave rising towards 4.0 volts is observed. This is entirely consistent with the peak oxidation potential, $E_{pa}$, of 0.96 volts vs. SSCE ($\sim$4.2 volts relative to lithium) observed for polythiophene in acetonitrile (12). This is anodic of peak oxidation potentials for $AlCl_4^-$ oxidation to $Cl_2$ and implies that $AlCl_4^-$ oxidation would occur before the polymer becomes oxidized.

The voltammogram of the poly-3-methylthiophene-coated electrode showed a principal reduction wave centered about 2.9 volts. This is primarily due to $SO_2$ reduction. Upon scanning anodic with the poly-3-methylthiophene film, a large oxidation wave at 3.9 volts was observed. This is believed to be due to polymer oxidation. This is entirely consistent with the peak oxidation potential, $E_{pa}$, of 0.72 volts vs. SSCE ($\sim$4.0 volts relative to lithium) observed for poly-3-methylthiophene in acetonitrile (12). Upon subsequently scanning anodic, a reasonably large reduction wave at 2.9 volts is observed again which suggests that the polymer oxidation wave at 3.9 volts is capable of chemically oxidizing the insoluble reduced product.

The lower peak oxidation potential, $E_{pa}$, in poly-3-methylthiophene can be explained if one considers the electron as a nucleophile. When the chemical group attached to the thiophene is an electron donator either by induction or resonance, the oxidation potential should be lowered. Groups such as $CH_3$ or $CH_2$ donate electrons inductively while N donates electrons by resonance. The addition of an electron-donating methyl group to the thiophene allows the oxidation potential of the poly-3-methylthiophene to be lowered to below 4.0 volts. Although the peak oxidation potential at which polythiophene oxidizes is considerably above 4.0 volts in $Li(SO_2)_3AlCl_4$, it is predictably lowered and occurs at $\sim$3.9 volts in the poly-3-methylthiophene. Thus, poly-3-methylthiophene alone or electropolymerized on carbon black surfaces promises to be a suitable polymer for use in rechargeable $Li/SO_2$ cells.

## 4. CONCLUSIONS

These studies have indicated that (i) polyacetylene can act as a catalytic electrode in an aqueous electrolyte for the reduction of gaseous oxygen, hydrogen peroxide or perchloric acid and (ii) poly-3-methylthiophene can act as the catalytic electrode for the reduction of $SO_2$ in $Li(SO_2)_3AlCl_4$ and permit the cathode discharge product to become rapidly oxidized without resorting to use of chlorine as an intermediate.

## 5. ACKNOWLEDGEMENTS

I wish to thank Professor Alan G. MacDiarmid for numerous technical suggestions and advice during the study of the electrocatalytic properties of polyacetylene in aqueous electrolytes. This work was supported by the US Office of Naval Research. I also with to thank Dr. Michael Binder and Dr. Sol Gilman for helpful advice and assistance during the study of polythiophene and poly-3-methylthiophene in inorganic electrolytes. This work was part of an Army internally funded research and development project (ILIR) on rechargeable lithium cells. Finally, I wish to thank Laura Thorsen who provided valuable manuscript assistance.

# 6. REFERENCES

1. J. M. Pochan, H. W. Gibson and F. C. Bailey, J. Polym. Sci., Polym. Lett. Ed., 18, 447 (1980); J. M. Pochan, D. F. Pochan, M. Rommelmann and H. W. Gibson, Macromolecules, 14, 110 (1981); J. M. Pochan, H. W. Gibson and J. Harbour, Polymer, 23, 439 (1982); H. W. Gibson and J. M. Pochan, Macromolecules, 15, 242 (1982).

2. R. J. Mammone and A. G. MacDiarmid, Synthetic Metals, 9, 143 (1984).

3. A. G. MacDiarmid, R. J. Mammone, N. L. D. Somasiri and J. R. Krawczyk in Energy Technology XI (R. F. Hill, ed.), Government Institutes, Inc., 1984, pp. 577-586.

4. W. Wanqun, R. J. Mammone and A. G. MacDiarmid, Synthetic Metals, 10, 235 (1985).

5. R. J. Mammone and A. G. MacDiarmid, J. Chem. Soc., Faraday Trans. I, 81, 105 (1985).

6. R. J. Mammone, Ph.D. Thesis, University of Pennsylvania, Philadelphia, PA (1985).

7. R. J. Nowak, B. Weiner and J. Calvert, The Electrochemical Society Extended Abstracts, Vol. 85-1, No. 89, Toronto, Ontario, May 12-17, 1985, pp. 131-132.

8. R. C. McDonald, W.-T. Wang, P. Cukor and M. F. Rubner (September 10, 1985), US Patent No. 4,540,641.

9. R. J. Mammone and M. Binder, J. Electrochem. Soc., 133, 1312 (1986).

10. R. J. Mammone and M. Binder, J. Electrochem. Soc. (accepted for publication).

11. A. N. Dey, H. C. Kuo, D. Foster, C. Schlaikjer and M. Kallianidis, Extended Abstracts of the 3rd International Meeting on Lithium Batteries, Kyoto, Japan, May 27-30, 1986, pp. 173-181.

12. R. J. Waltham, J. Bargon and A. F. Diaz, J. Phys. Chem., 87, 1459 (1983).

13. H. Shirakawa and S. Ikeda, Polym. J., 2, 231 (1971); H. Shirakawa, T. Ito and S. Ikeda, J. Polym. Sci., Polym. Chem. Ed., 12, 11 (1974); ibid., 13, 1943 (1975); C. K. Chiang, Y. W. Park, A. J. Heeger, H. Shirakawa, E. J. Louis and A. G. MacDiarmid, J. Chem. Phys., 69, 5098 (1978).

# POLYPHTHALOCYANINES

Michael Hanack*, Sonja Deger, Uwe Keppeler, Armin Lange,
Andreas Leverenz and Manfred Rein
Institut für Organische Chemie, Lehrstuhl für Organische Chemie II der
Universität Tübingen, Auf der Morgenstelle 18, D-7400 Tübingen,
West-Germany

ABSTRACT. Bridged macrocyclic transition metal complexes using phthalocyanine and naphthalocyanine as macrocycles e.g. [PcML]$_n$ with M = Fe,
Ru Co and L = dib, tz, CN show good semi-conducting properties without
external doping. The mechanism of the formation of the bridged macrocyclic metal complexes is studied by $^1$H-NMR spectroscopy using the
monomer PcFe(Me$_4$dib)$_2$.
    Phthalocyaninatoiron and ruthenium (PcFe, PcRu) as well as
2,3-naphthalocyaninatoiron (2,3-NcFe) form s-tetrazine bridged compounds e.g. [PcRu(tz)]$_n$, which exhibit high semi-conducting properties
without additional doping. Their conductivities are comparable to
cyanide bridged metallomacrocycles [PcMCN]$_n$.
    Electrochemical doping of bridged phthalocyaninato transition
metal complexes [PcML]$_n$ is reported.

## INTRODUCTION

Transition metal complexes, which are linked by linear bridging
ligands L containing delocalizable $\pi$-electrons to form polymeric
stacked arrangements, have been synthesized by us using either phthalocyanine (Pc) or tetrabenzoporphyrine (TBP) as macrocycle, Fe, Ru, Co,
Rh and others as central metals and e.g. pyrazine (pyz), diisocyanobenzene (dib) and cyanide (CN⁻) as bridging ligands. This type of
polymers can be used with and without doping as good semi-conducting
materials [1].

## RESULTS

The characterization of the semi-conducting bridged phthalocyaninato-
metal polymers up to now was mostly done by IR spectroscopy or thermogravimetric methods. We report here about NMR and Mößbauer studies to

*L. Alcácer (ed.), Conducting Polymers, 173–178.*
© *1987 by D. Reidel Publishing Company.*

investigate the mechanism of formation of the bridged polymers. For this purpose monomeric and bridged phthalocyaninatoiron(II)- and ruthenium(II) compounds with 2,3,5,6-tetramethyldiisocyanobenzene ($Me_4dib$) were synthesized.

The reaction of $Me_4dib$ with phthalocyaninatoiron(II) and ruthenium(II) (PcFe, PcRu) yields, depending on the reaction conditions bisaxially coordinated monomeric derivatives of the stoichiometry $PcM(Me_4dib)_2$; M = Fe, Ru, or one-dimensional polymeric chain structures of the formula $[PcM(Me_4dib)]_n$, M = Fe, Ru.

The monomeric compounds $PcM(Me_4dib)_2$, M = Fe (1), M = Ru (2) are characterized through their $^1H$-NMR spectra. If the $^1H$-NMR spectrum of 1 and 2 is recorded again after thirty minutes and several time subsequently at a later time, beside the original spectrum of the monomer 1 and 2, two groups of signals at a higher field are observed, which increase their intensity the longer the monomer stays in solution. For the peripheric phthalocyaninato-protons an upfield shift of about 0.4 ppm is observed, which definitely shows the formation of dimers, trimers and oligomers. The polymerization can also be followed by observing the upfield shift of the methyl groups of the axial ligands in 1 and 2. From the $^1H$-NMR investigations the mechanism of the formation of the oligomers as shown in scheme 1 can be derived.

SCHEME 1

| | | | | |
|---|---|---|---|---|
| I   m PcM | + 2m L | m $PcML_2$ | | |
| II  m $PcML_2$ | | m [PcML] | | + m L |

| | | | | |
|---|---|---|---|---|
| III m [PcML] | + $PcML_2$ | m L-PcM-L-PcM-L | + m L |
| IV  m  L-PcM-L-PcM-L | | m [L-PcM-L-PcM] | + m L |
| V   m  [L-PcM-L-PcM] | + $PcML_2$ | m L-PcM-L-PcM-L-PcM-L | + m L |

Mößbauer spectroscopy is another important method to obtain additional information about the structure of the bridged phthalocyaninatoiron compounds. The isomer shift ($\delta$) and the quadrupol splitting ($\Delta E_Q$) have been measured for the monomer 1 ($\delta$ = 0.11 mm/s, $\Delta E_Q$ = 0.65 mm/s). The data of $[PcFe(Me_4dib)]_n$ $\delta$ = 0.$\overline{1}$4 mm/s and $\Delta E_Q$ = 0.89 mm/s are another definite proof for the hexacoordination of the dib-bridged compounds.

Depending upon the bridging ligands, e.g. pyrazine or diisocyanobenzene whole the variety of bridged macrocyclic metal complexes e.g. $[PcFe(pyz)]_n$ and $[PcFe(dib)]_n$ [1,2] have been synthesized and characterized. These compounds show a semi-conducting behavior to an upper limit of $10^{-5}$ S/cm without oxidative doping.

The electrical properties can be highly improved if a bridging ligand containing more nitrogen atoms and thereby lowering the energy of the LUMO [3] is used. A good example for this bridging ligand is s-tetrazine (tz), which can be used to form thermally stable bridged macrocyclic metal complexes e.g. $[PcFe(tz)]_n$.

The synthesis of the already reported[4] $[PcFe(tz)]_n$ is carried out by heating PcFe in a solution of s-tetrazine in chlorobenzene at 100°C for 24 h. The product of the heterogeneous reaction is a thermal- and light-stable blue powder. The stoichiometry and purity of the compound is well established by different methods such as IR-, far-IR-, Raman-, UV/Vis-spectroscopy, microanalysis and combined thermogravi-metric and differential thermal analysis. $[PcFe(tz)]_n$ splits off s-tetrazine above 125°C with an exothermic maximum at 310°C, PcFe remaining as residue. The synthesis of a complex $PcFe(tz)_2$, containing two bisaxial coordinated molecules of s-tetrazine, is possible in a heterogeneous reaction of PcFe with s-tetrazine in ethanol.

For $PcFe(tz)_2$ and $[PcFe(tz)]_n$, as in the case of $PcFe(pyz)_2$ and $[PcFe(pyz)]_n$ [5], the Mößbauer spectra show a decrease of $\Delta E_Q$ relative to $\alpha$-PcFe. $\beta$-PcFe: $\delta$ = 0.383 mm/s, $\Delta E_Q$ [mm/s] 2.600; $PcFe(tz)_2$: $\delta$ = 0.153, $\Delta E_Q$ 1.79 mm/s; $[PcFe(tz)]_n$: $\delta$ = 0.127, $\Delta E_Q$ 2.274; relative to metallic iron. This well known effect is caused by the hexa-coordinated iron in contrast to the square planar iron in PcFe.

The complexes $PcRu(tz)_2$ and $[PcRu(tz)]_n$ [6] were also prepared. The monomer $PcRu(tz)_2$ is synthesized by stirring a solution of s-tetra-zine in chloroform with PcRu for 1 h at 70°C. However, when the reac-tion time is extended to 70 h a mixture of $PcRu(tz)_2$, soluble oligo-mers and $[PcRu(tz)]_n$ is obtained. Extraction of this mixture with $CHCl_3$ yields pure $[PcRu(tz)]_n$.

For the first time 2,3-naphthalocyanine (2,3-Nc) is used as the macrocycle. 2,3-Naphthalocyaninato iron can be prepared by the drop-wise addition of $Fe(CO)_5$ to a solution of 2,3-dicyanonaphthalene in 1-chloronaphthalene at 250°C in the absence of oxygene.

Reacting 2,3-NcFe with s-tetrazine (tz) yields the corresponding bridged 2,3-NcFe compound $[2,3-NcFe(tz)]_n$. Investigation of $[2,3-NcFe(tz)]_n$ by Mößbauer spectroscopy permits the distinction between a polymeric, hexacoordinated and a pentacoordinated structure of this compound. The isomer shift and the quadrupol splitting are in the same range as in $[PcFe(tz)]_n$. The isomer shift $\delta$ of $[2,3-NcFe(tz)]_n$ at 0.17 mm/s and the quadrupol splitting $\Delta E_Q$ at 1.95 mm/s differs from 2,3-NcFe with $\delta$ = 0.36 mm/s and $\Delta E_Q$ = 2.20 mm/s. This values give evidence for a hexacoordinated iron and confirm the bridged structure of $[2,3-NcFe(tz)]_n$, because pentacoordinated high spin iron compounds show a higher quadrupol splitting in contrast to the square planar coordinated iron [7].

The s-tetrazine bridged macrocyclic compounds show in general comparatively high conductivities (Table 1) without additional doping.

TABLE 1

Conductivity data of s-tetrazine bridged iron macrocycles

| Compound | $\sigma_{RT}$ [ S/cm] | $E_a$ [eV] |
|---|---|---|
| PcFe(tz)$_2$ | $< 10^{-9}$ [a] | - |
| [PcFe(tz)]$_n$ | $2 \cdot 10^{-2}$ [b] | 0.10 |
| PcRu(tz)$_2$ | $< 10^{-11}$ [a] | - |
| [PcRu(tz)]$_n$ | $1 \cdot 10^{-2}$ [b] | 0.10 |
| [2,3-NcFe(tz)]$_n$ | $3 \cdot 10^{-1}$ [b] | 0.07 |

[a]Two-probe-technique, [b]four-probe-technique.

The reason for the high conductivities of the s-tetrazine bridged metallomacrocycles is not yet completely understood. The electronic structure of [PcFe(pyz)]$_n$ has been studied by means of the tight-binding (LCAO) method. As a result a band gap of about 0.7 eV and semi-conducting behaviour was predicted [3]. The band gap according to these calculations is mostly determined by the difference in energies between the LUMO of the bridging ligand and the HOMO of the transition metal d$_{xy}$ orbital. The higher conductivity of the s-tetrazine bridged compounds [PcFe(tz)]$_n$, [PcRu(tz)]$_n$ and [2,3-NcFe(tz)]$_n$ is thereby explainable by the low lying LUMO of this ligand.

A second class of bridged macrocyclic metal complexes with comparatively high semi-conducting behaviour is created by cyanide bridged cobalt and iron phthalocyanine and tetrabenzoporphyrine (TBP) compounds. [PcCoCN]$_n$ is well investigated [8] and preparation methods have been extended to the analoguous [TBPCoCN]$_n$ [9] and [PcCoSCN]$_n$ [10] which all show comparable electrical properties (Table 2).

TABLE 2

Conductivity data of cyano- and thiocyanato-bridged cobaltmacrocycles

| Compound | $\sigma_{RT}$[S/cm][a] | $E_a$ [ eV] |
|---|---|---|
| [PcCoCN]$_n$ | $2 \cdot 10^{-2}$ | 0.10 |
| [TBPCoCN]$_n$ | $4 \cdot 10^{-2}$ | 0.11 |
| [PcCoSCN]$_n$ | $6 \cdot 10^{-3}$ | 0.22 |

[a]Four-probe-technique.

$PcCoCN_n$ shows unique properties e.g. it can be dissolved in conc. $H_2SO_4$ and $CF_3SO_3H$. From sulfuric acid solutions of $PcCoCN_n$ containing the e.g. polyamide Kevlar flexible fibres can be spun which show good antistatic properties without doping.

## Electrochemical doping of $[PcFe(pyz)]_n$

The rather low semi-conducting behavior of $[PcFeL]_n$ compounds (L = pyz, dib) can be increased not only by oxidative doping with iodine as reported earlier [5]; systematic investigations on electrochemical doping processes are carried out. As an example electrochemical doping of $[PcFe(pyz)]_n$ ($\sigma_{RT} = 2 \cdot 10^{-6}$) with a variety of counterions $BF_4^-$, $B(Ph)_4^-$, $PF_6^-$ and $ClO_4^-$ is described (Table 3). The doping experiments are done in $CH_2Cl_2$ or $H_2O$ with current densities about 90-200 $\mu A/cm^2$ and a 0.1 mol/l solution of the supporting electrolytes (except $(Bu_4N)B(Ph)_4$ : 0.05 mol/l). The electrolyses is stopped between 0.25 and 1.0 F/mol (of constitutive unit of the polymer); best conductivities were reached at 0.5 F/mol.

TABLE 3

Electrochemical doping of $[PcFe(pyz)]_n$ using a current density of 120 $\mu A/cm^2$

| Supporting electrolyte | Solvent | $\sigma_{RT}$ [S/cm][a] |
|---|---|---|
| $(Bu_4N)BF_4$ | $CH_2Cl_2$ | $4 \cdot 10^{-2}$ |
| $(Bu_4N)PF_6$ | $CH_2Cl_2$ | $3 \cdot 10^{-2}$ |
| $(Bu_4N)B(Ph)_4$ | $CH_2Cl_2$ | $2 \cdot 10^{-6}$ |
| $LiClO_4$ | $H_2O$ | $3 \cdot 10^{-3}$ |

[a]Four-probe-technique.

Doping experiments by a two electrode constant potential method lead to similar results.

IR spectra of the doped species show a broad electronic absorption.

## ACKNOWLEDGEMENTS

This work was supported by the Stiftung Volkswagenwerk, the Bundesministerium für Forschung und Technologie and the "Forschungsschwerpunkt Nr. 39, Neue Materialien" at the University of Tübingen.

REFERENCES

1  a) M. Hanack, Chimia, 37 (1983) 238; b) M. Hanack, A. Datz,
   R. Fay, K. Fischer, U. Keppeler, J. Koch, J. Metz, M. Mezger,
   O. Schneider and H.-J. Schulze in T. Skotheim (ed.), Handbook
   on Conducting Polymers, Marcel Dekker, New York 1985.
2  O. Schneider and M. Hanack, Chem. Ber., 116 (1983) 2088.
3  E. Canadell and S. Alvarez, Inorg. Chem., 23 (1984) 573.
4  O. Schneider and M. Hanack, Angew. Chem., 95 (1983) 804.
5  B.N. Diel, T. Inabe, K.K. Jaggi, J.W. Lyding, O. Schneider,
   M. Hanack, C.R. Kannewurf, T.J. Marks and C.H. Schwartz, J. Am.
   Chem. Soc., 106 (1984) 3207.
6  U. Keppeler, Ph. D. Thesis, University of Tübingen, Germany, 1985.
7  G. Silver, B. Lukas and G. Al-Jaff, Inorg. Chim. Acta, 91 (1984)
   125.
8  J. Metz and M. Hanack, J. Am. Chem. Soc., 105 (1983) 828.
9  M. Hanack and C. Hedtmann-Rein, Z. Naturforsch., 40b (1985) 1087.
10 M. Hanack and C. Hedtmann-Rein, Inorg. Chem., in print.

LANGMUIR–BLODGETT DEPOSITION OF AMPHIPATHIC AZOBENZENE COMPOUNDS FOR
SURFACE ACTIVATION AND FABRICATION OF FUNCTIONALISED THIN FILMS

C. G. Morgan, Y. P. Yianni and S. S. Sandhu
University of Salford
Department of Biological Sciences
Salford M5 4WT
UK

ABSTRACT.  Langmuir–Blodgett techniques provide one means of preparing
thin films ordered in two dimensions.  An alternative approach uses self
organisation of reactive molecules which can chemisorb onto surfaces.
Each technique has advantages and limitations, especially in the context
of preparation of functionalised surfaces such as conducting films.
Recent work has produced a group of amphipathic azobenzene derivatives
which are designed to allow either chemisorption or physisorption onto
surfaces, and which should allow hybrid film fabrication, combining advantages of each technique.   The synthesis and surface properties of
these molecules is discussed, and application areas explored, particularly in the context of surface functionalisation.   Successful exploi-
tation of thin film technology in molecular electronics requires that
organisation be controllable in the plane of a given monolayer, as well
as from one monolayer to another, and this is discussed in relation to
the azobenzene derivatives mentioned.

Conductive polymers have very considerable potential in many fields, but
among the most interesting are the, as yet, rather speculative areas
broadly classified as 'molecular electronics'.
    The fabrication of useful and reliable electronic devices based on
organic molecules or biological macromolecules offers formidable chal-
lenges.  A central problem, common to many diverse disciplines, is the
achievement of controlled molecular organisation, and it is this aspect
which we wish to discuss here.   Control of atomic or molecular organisation in three dimensions is an extremely difficult problem, as yet unsolved.   Molecular engineering has progressed in the areas of solution-
and interface chemistry and model catalysts having considerable specificity have been designed based on our increasing understanding of struc-
ture/function relationships in enzymes and similar macromolecules.   We
are, however, a very long way from a true understanding of molecular
self assembly, and how biological materials achieve their form and function.

L. Alcácer (ed.), Conducting Polymers, 179–187.

To date, most experimental studies of controlled molecular assembly have been conducted on molecular monolayers or multilayers prepared by deposition using 'Langmuir-Blodgett' techniques(1).

Such 'L-B' films are prepared by transfer of monolayers of an amphipathic substance from an interface onto a substrate, usually by dipping the substrate through the interfacial film.  In order for a monolayer to effectively transfer, it is usually necessary to impose a surface pressure sufficient to orient and align the film-forming molecules at the interface.  In suitable cases, the application of surface pressure can overcome steric constraints to molecular close-packing and favourably orient molecular features.  Consequently, Langmuir-Blodgett films can be formed from a wide variety of amphipathic molecules, including many lipids and macromolecules.  Langmuir-Blodgett films can be sequentially deposited to form multilayers and these can be heterogeneous since the interfacial film can be varied from one transfer to the next.  Such composite L-B films offer one rather primitive, but nonetheless extremely interesting and useful, example of 'molecular engineering'.

A host of extremely ingenious and informative experiments has been conducted on multilayer L-B films, including studies of fluorescence energy transfer, dye sensitisation processes, electron tunnelling and many others(2).

The Langmuir-Blodgett technique is not the only way to prepare monomolecular films on substrates.

In suitable circumstances a surface monolayer will spontaneously adsorb from a solution of amphipathic molecules in a partially ordered solvent.  A very interesting application of spontaneous adsorption was reported fairly recently by Jacob Sagiv of the Weizmann Institute(3). Sagiv prepared films on hydrophilic substrates by exposure to a solution of a terminally-unsaturated alkylchlorosilane in a hexadecane based solvent.  A surface monolayer rapidly forms by chemisorption through formation of oxygen-silicon linkages to the substrate.  The novel aspect of this work was the ability to subsequently add a second monolayer of the same or different silane via a chemical activation process.  The terminal double bonds of the molecules of the monolayer were converted through hydroboration and alkaline peroxide oxidation into terminal alcohols.  The chemical reactions are rapid, mild and essentially quantitative and the resulting surface is hydrophilic.  Further chemisorption can add a second monolayer and the process could, in principle, be repeated as many times as desired.  In practice there are difficulties, since film quality deteriorates if multilayer fabrication is attempted. This is not surprising, since incomplete activation and packing imperfections will eventually limit film properties.

Chemisorption and L-B deposition are complementary in many ways. Chemisorption is rapid and requires no special equipment.  The surface films are inherently anchored and can be formed on powders or rough surfaces.  The range of film-forming molecules is limited however and steric constraints, such as might result from attempts to functionalise alkyl chains, are extremely important to film quality.  Such functionalisation would be necessary if chemisorbed <u>conductive</u> films were to be fabricated and it remains to be seen whether this proves possible.  On

the other hand, L-B films can be prepared from a wide variety of molecules since surface pressure induces alignment and overcomes steric res-
trictions. The deposition process can be repeated many times to form
multilayers, but these layers are not anchored other than by physical
adsorption.    Anchoring and cross-linkage can be achieved chemically
after film formation (e.g. by polymerisation), but film quality can be
affected by slight volume changes during such processing.    Langmuir-
Blodgett techniques require care and deposition rates can be low, but
there is no practical alternative in many cases.

Any practical thin film 'molecular electronic' device will need to
have a significant lifetime.  If such a device has features fabricated
in the plane of the molecular film, diffusion will tend to randomise the
surface unless this is physically anchored or cross-linked.    Although
diffusion on adsorbed monolayers is slow, it is not negligible. The root
mean square displacement $(\bar{X})$ of a molecule diffusing in two dimensions
is given by $\bar{X} = (4Dt)^{\frac{1}{2}}$, where D is the lateral diffusion coefficient and
t is time.  If one assumes, for example, a diffusion coefficient of $10^{-10}$
cm.sec$^{-1}$ a molecule will diffuse an 'average' distance of 20 $\mu$m in only
$10^4$ seconds.  If we consider a device with submicron features and a projected lifetime of months or years, the need to reduce diffusion to negligible proportions becomes obvious.    A related problem is the possibility of diffusion between adjacent monolayers.    Again, this is a very
slow process but cannot be ignored for a practical device.    These considerations suggest that any 'molecular electronic' device of significant complexity must be somehow cross-linked or otherwise polymerised.
It should be designed so that functional groups can be selectively incorporated in the structure and the pattern of incorporation should be
controllable.

The fabrication of patterns on monolayer surfaces might be achieved
in several ways.  The surface could be modified chemically or physically
by selective ion or electron bombardment, or by photolithographic methods.  The modification could alter the pattern of subsequent deposition
of further monolayers (e.g. by rendering a hydrophobic surface locally
hydrophilic or vice versa), or might produce chemical activation or de-
activation towards a surface reaction.

We have approached the goal of functionalised monolayer fabrication
by synthesis of a group of azobenzene-containing amphipathic molecules.
Such molecules can be deposited by either chemisorption or physisorption
and should allow hybrid films to be fabricated.    The monolayers are reversibly photochromic (which might prove useful in sensor fabrication)
and are capable of chemical activation for surface attachment of macromolecules, etc.    In addition, the activated surface can be photochemi-
cally modified so that photolithography could be used to fabricate surface structures such as domains of particular composition.    With particular regard to conductive polyers, polyanilines could be locally formed
as discussed later.

Long chain azobenzene-containing acids can be synthesised fairly
easily and in good yield from readily available precursors(4).  The synthesis of a short chain azobenzene-containing acid is shown in Figure
1(a).    The azobenzene unit does not inhibit formation of Langmuir-

Blodgett films from this acid and a typical isotherm is shown in Figure
1(b).    Films formed from this acid are transferable and multilayer deposition has been demonstrated in preliminary experiments.    The molecular area for the acid measured from the isotherm is in good accord with
that calculated from molecular models and the film is close packed.    The

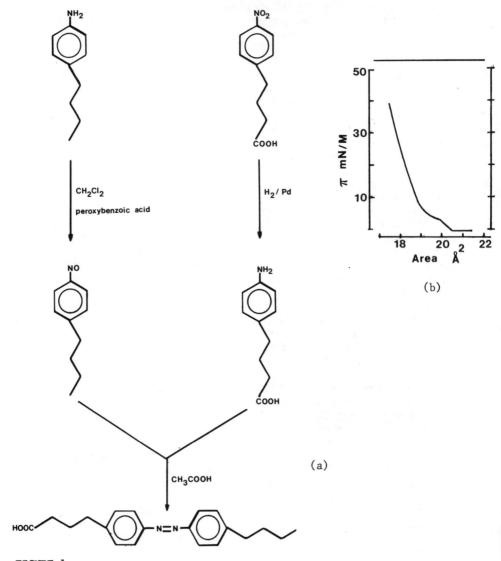

(b)

(a)

FIGURE 1

The synthesis of 4(4-n.butylphenylazo)-phenylbutyric acid is shown in
I(a).    A typical pressure/area isotherm for the acid is shown in I(b)
for a pure water subphase at pH 6, 20°C.

azobenzene-containing acid has been incorporated into a group of phospholipids which also show excellent surface behaviour(5). The structures
of two of these, which we have called 'palmitoyl Azo PC' and 'Bis-Azo
PC' (since both are phosphatidylcholines) are shown in Figure 2(a) and
molecular models are shown in Figure 2(b).    Bis-Azo PC is especially
interesting, since it behaves in many ways like dipalmitoylphosphatidylcholine (which it resembles in size).    Bis-Azo PC forms lipid vesicles
on dispersion in water and these show co-operative phase behaviour, having a marked phase transition at around 42°C.

(1)

(2)

(3)                                              (4)

FIGURE 2

The structures of azobenzene-containing phospholipids 'palmitoyl Azo PC'
(1) and 'Biz-Azo PC' (2).    Molecular models of the structures are shown
in (3) ('palmitoyl Azo PC') and (4) ('Bis-Azo PC'), where the azobenzene
unit is in the _trans_ form.

Surface Pressure-Area isotherms for both phosphlipids are shown in
Figure 3 for the all-trans molecules and also for the photostationary
state mixture after UV photoisomerisation.    As expected, isomerisation
increases the apparent molecular area, since the 'cis' azobenzene isomer
is a non-linear molecule and causes a 'kink' in the lipid acyl chain
which affects the packing properties of the film.

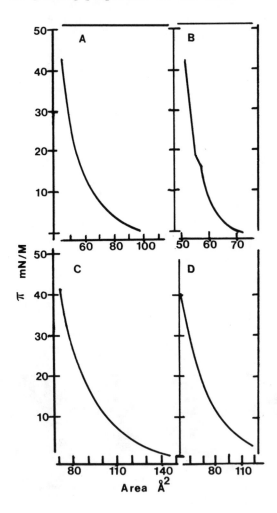

FIGURE 3

Pressure/area isotherms for 'palmitoyl Azo PC' (A) in the trans form and
(C) in the 'cis' form.    Isotherms for 'Bis-Azo PC' in the trans form
(B) and 'cis' form (D) are also shown.

'Cis' form refers to the photostationary state on illumination at 360nm.
Isotherms are on pure water at 20°C, pH 6.

Deposited monolayer films of the azolipids are not easily affected by sodium dithionite, which rapidly cleaves azobenzene linkages in most compounds.   Vesicles of Bis-Azo PC are similarly resistant.   This presumably reflects the close-packed nature of the structures, which prevents access of reductant to the Azo-linkage.   Nascent hydrogen from slightly acidified borohydride solution rapidly reduces the compounds to amines, cleaving the azobenzene linkage.   It is likely, though not yet demonstrated, that photoisomerisation of the azobenzene linkage will markedly affect the rate of reductive cleavage by mild reagents.   The cis-azobenzene unit is more polar than the trans form and the isomerisation locally disorders molecular packing.   This aspect might allow selective cleavage in irradiated areas and is to be investigated.

Cleavage of azobenzene units produces a surface film of aromatic amine.   The surface should provide a very useful experimental means to study the formation of polyanilines on oriented monolayer surfaces.

Constraints on orientation will affect the formation and properties of polyaniline from the substituted aniline surface groups. In addition, the surface is well suited to coating by 'graft' polymerisation where the surface aromatic amines anchor a conventional polyaniline film to the monolayer.

Surface aromatic amines can be diazotised and used to bind enzymes, proteins or other macromolecules to the surface.   Similar technology is presently used to activate glass surfaces for enzyme immobilisation. Aromatic diazonium compounds are photosensitive and have long been used in 'diazotype' printing.   It follows that diazotisation of a monolayer surface followed by exposure through a mask allows reactive groups to be selectively destroyed in illuminated regions. This offers the possibility of selective surface modification, by binding enzymes or other functional units to specific regions of the monolayer.   Alternatively, a diazotised surface can be efficiently converted to aromatic azide which is chemically stable, but a well known photoactivated nitrene precursor.

The requirement for anchoring of L-B films can be achieved by the use of phospholipids bearing photoactivatable head groups, which covalently bind to surfaces on illumination.   For glass or silicon substrates the surface must be first coated with an organic layer using, for example, 3-aminopropyl triethoxysilane or similar material.

The 4(4'-n.butyl phenylazo)phenylbutyric acid described earlier is easily converted to an acid chloride and thence to a wide variety of activated or activatable species, including acylazide, amine (by acyl azide rearrangement), diazoketone (with diazomethane), iodide (by Hunsdiecker reaction) and alkene (by treatment with Wilkinson's catalyst). Alkenes are easily converted to silanes, so that this offers a direct route to a derivative of the type used by Sagiv in the work mentioned earlier.

In principle, the acid chloride could be reacted with an amino-labelled surface to give an amide.   Cleavage of the azobenzene group then yields a new amine for a further round of reaction.   This scheme would give an anchored multilayer of linearly polymerised molecules. However, although the reaction undoubtedly would proceed, the azobenzene containing acid described is unlikely to spontaneously form a close packed chemisorbed film.   Molecules which form such films have long

alkyl chains which spontaneously interact through short range forces in
the partially ordered hydrocarbon solvents used.   It was therefore
thought desirable to synthesise azobenzene derivatives having long alkyl
chains and to study their surface properties.   A typical synthesis
scheme starts with a long chain fatty acid, where one double bond is
present at a defined position.   Oxidative cleavage of this generates a
terminal aldehyde group, which is reacted with 4-nitrobenzyl triphenylphosphonium bromide in a Wittig reaction.   Catalytic hydrogenation of
the product is followed by coupling with a nitroso compound, itself prepared by low temperature oxidation of an appropriate aromatic amine.   To
date we have prepared a variety of these acids of widely varying chain
length, where the position of the azobenzene group is varied within the
chain.   The surface behaviour of a typical long chain acid is shown in
Figure 4(a).   All of the acids tested so far have formed surface films.

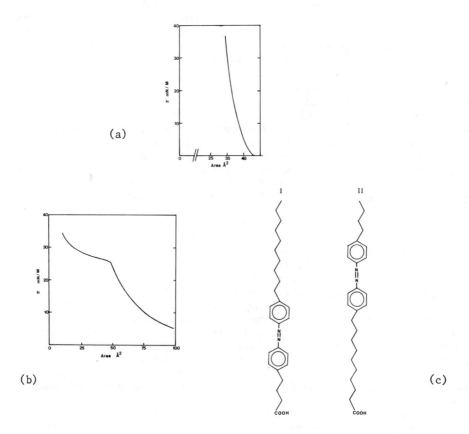

## FIGURE 4

Isotherm of a 'typical' long chain azobenzene acid [structure shown in
4(c) I] is shown in 4(a).   Anomolous results are found for compound
shown in 4(c) II, the isotherm of which is shown in 4(b).

One of the acids [isotherm shown in Figure 4(b)] shows very unusual behaviour, in that at low surface pressure the molecule is clearly oriented parallel to the aqueous surface. On increasing the pressure, the isotherm shows an apparent limiting surface area which is physically impossible given the molecular size. It appears that either the monolayer is in equilibrium with dissolved material (unlikely, given the molecular weight) or that multilayer films form in some areas. The isotherm shows ready reversibility as the film is expanded, though on extreme compression the film will collapse. The structures of an acid giving a 'typical' isotherm and the anomolous molecule are shown in Figure 4(c). The possibility of multilayer film formation is a particularly interesting one, because such films are not forbidden thermodynamically, but their existence is not unanimously accepted.

Experiments are presently underway to study spontaneous adsorption of the long chain azobenzene amphiphiles. It is clear from the surface measurements that the azobenzene unit offers no significant steric constraint to such film formation.

To summarise, azobenzene derivatives can be used to form high quality monolayer films. The azo-linkage provides a 'masked' functionality which is suited to a wide range of chemical and photochemical manipulations. In particular, it offers a means to fabricate patterns of adsorbed species in the plane of the monolayer, using photolithography. These aspects are of considerable importance to the development of biosensors and molecular electronic devices and to the fabrication of ordered conductive films.

REFERENCES

1.  For a review see Gaines, G. L. 'Insoluble monolayers at liquid–gas interfaces'. NY; London: Intersciences (1966)

2.  Kuhn, H. (1983) Thin Solid Films, **99**, 1–16

3.  Netzer, L., Iscovici, R. and Sagiv, J. (1983) Thin Solid Films, **99** (1–3), 235–241

4.  Morgan, C. G., Thomas, E. W., Yianni, Y. P. and Sandhu, S. S. (1985) Biochim. Biophys. Acta, **820**, 107–114

5.  Sandhu, S. S., Yianni, Y. P., Morgan, C. G., Taylor, D. M. and Zaba, B. N. (1986) Biochim. Biophys. Acta (in press)

# ELECTRONICALLY CONDUCTING POLYMER GAS SENSORS

J.J. Miasik[+], A. Hooper, P.T. Moseley and B.C. Tofield
Materials Development Division,
United Kingdom Atomic Energy Authority,
Harwell Laboratory,
Oxfordshire,  OX11 ORA,
United Kingdom.
[+]Present Address:  Crystalox Ltd.,
 1, Limborough Road,
 Wantage,
 Oxfordshire, United Kingdom.

ABSTRACT.  It has been shown that devices can be fabricated using electronically conducting organic polymers for the ambient temperature detection of several industrially important gases.  In particular the resistance of thin films of polypyrrole has been shown to increase in the presence of 0.1% ammonia in air and to decrease in the presence of 0.1% nitrogen dioxide and 0.1% hydrogen sulphide.  Devices based on conducting polymers may thus offer advantages in environmental monitoring over presently available semiconductor sensors which generally operate at elevated temperatures.

## 1.   INTRODUCTION

In recent years there has been a growing interest amongst industrialized societies in the development of sensors for monitoring the composition of gaseous ambients.  The motivation for this interest derives both from an enthusiasm for the increased economies that are achievable by tighter control of processes such as combustion and from a growing concern for environmental protection and safety.  The requisite electronics and signal processing capability for such applications are already to hand and the development of suitable sensors is seen as the major factor limiting market development (1).

Solid state gas sensors are already well established for a number of purposes (1), which are listed in Table I, but all the device types shown depend on sensor elements operating at elevated temperatures.  It would be a great advantage for sensors to operate entirely at ambient temperature since their power requirements could be met by small battery packs and the instruments would then be readily portable.

Aside from its temperature of operation the tin dioxide sensor which functions as a gas-sensitive resistor (2) offers a number of

189

L. Alcácer (ed.), Conducting Polymers, 189–198.
© 1987 by D. Reidel Publishing Company.

TABLE I    Solid State Gas Sensors

| Type | Purpose | Approximate Temperature of Operation (°C) |
|---|---|---|
| Galvanic Solid Electrolyte | Oxygen level in automobile exhausts and boiler flues. | 600 |
| Catalytic (pellistor) | Flammable vapours in oil and gas industries. | 500 |
| Semiconductor (a)  Bulk Conductivity ($TiO_2$) | Oxygen in automobile exhausts. | 800 |
| (b)  Surface Conductivity ($SnO_2$) | Toxic gases (CO, $H_2S$). | 400 |

desirable attributes (simplicity, ruggedness, cheapness, etc.) that ought to be retained in any viable alternative system.

Organic materials such as the metal phthalocyanines have been developed to give an extremely sensitive response to nitrogen dioxide (3) with only a minor interference from halogens, but even these materials operate at around 200°C and at present they show little promise of being developed for the detection of other toxic gases.

Conducting organic polymers have received much attention recently with considerable effort devoted to exploiting their utility in the development of high energy density secondary batteries. It is widely recognised that there will be other applications of these exciting new materials and if they can be prepared in an appropriate form and with sufficient conductivity then room temperature gas sensors may be one of these. It has been shown (4) that the resistance of polypyrrole deposited on a filter paper is sensitive to the presence of ammonia at room temperatue. The present paper describes the fabrication of thin films of polypyrrole (Fig. 1) and the sensitivity of the material to the presence of small concentrations of toxic gases in air. Some of the data have been briefly described elsewhere (5).

## 2.   MATERIALS AND METHODS

Thin films of conducting polymer were set down by a process of electro-polymerisation from aqueous solutions in which were dissolved the monomer (typically 0.06 molar) and a lithium salt of a suitable counter ion (generally 0.1 molar $LiBF_4$ was used). A DC potential of around

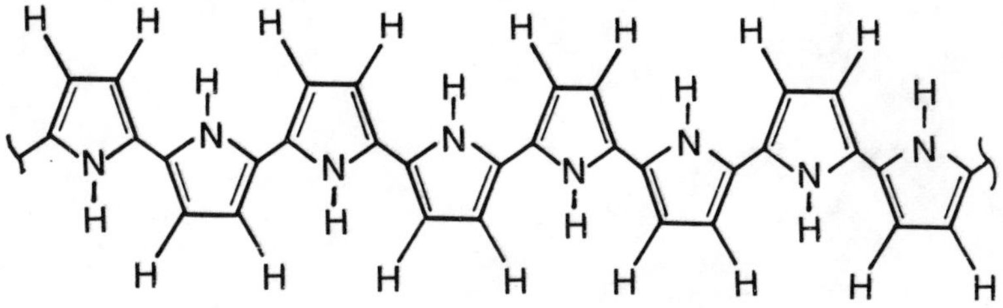

Figure 1.  Structure of polypyrrole.  High conductivity may be induced by negative ion doping to create positive defects on the polymer backbone.

5 volts was applied between the sensor (a screen printed gold interdigitated electrode array on an alumina tile) as anode and a gold wire as cathode.  Almost immediately the polypyrrole film became visible as a black film on the gold electrodes of the sensor (Fig. 2) and hydrogen was evolved at the cathode.  The fine electrode spacing allowed growth of the polymer film across the interelectrode gap.  The conductance of thin film sensors prepared in this way was adequate at ambient temperature for the resistance monitoring of gas sensor responses to be a straightforward matter.  In the form in which they were first prepared films were sensitive to electron-donating gases such as ammonia.  It was found that a post-fabrication treatment enabled the sensors to respond to additional gases.

Sensors like that shown in Fig. 2 were exposed to a flowing atmosphere of air containing the test gas in a glass apparatus in which changes in atmosphere were effected by computer directed switching of solenoid valves.  DC resistivity measurements were made with a digital multimeter and AC resistivity measurements (at 100 Hz, 1 kHz and 10 kHz) were made using a Hewlett-Packard LCR meter (model 4262A).

Gas sensitivity tests were performed on the blank interdigitated electrode arrays and it was confirmed that in the absence of a conducting polymer film there was no gas response.

The electronically prepared conducting polymers were amorphous and the assumption was made that the material underwent no chemical change during polymerisation beyond the incorporation of the tetrafluoroborate counter ion.

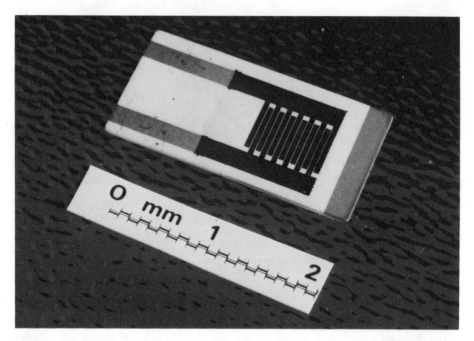

Figure 2. Gas sensor consisting of doped polypyrrole electrochemically polymerised on to interdigitated gold electrodes screen-printed on an alumina substrate.


3.   RESULTS

A preliminary qualitative test on a post-treated sensor (Fig. 3) shows that polypyrrole behaves as a quasi 'p' type material, offering resistance increases in the presence of a reducing gas (ammonia) and resistance decreases in the presence of an oxidising gas (nitrogen dioxide).   These results are consistent with a mechanism parallel to that thought to operate on tin dioxide in which gas molecules cause changes in the near surface charge carrier (here electron holes) density by reacting with surface adsorbed oxygen ions.   The quantitative response of a polypyrrole sensor to eight 15 minute pulses of 0.1% ammonia in air is shown in Fig. 4.   The size of the response to ammonia (a resistance increase) was sustained throughout this period but the background resistance rose during this time by around 20% of its initial value.   This rise in air-only resistance might be associated with changes in the moisture content of the polymer in which case stability could probably be achieved by modification of the sensor microstructure.

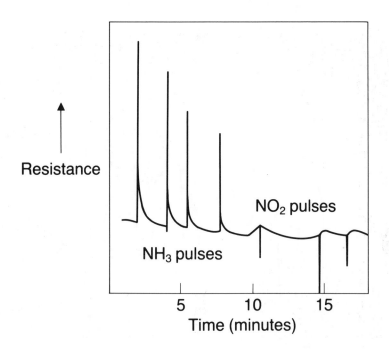

Figure 3. Qualitative DC resistance response characteristic at room
temperature of a conducting polypyrrole sensor to pulses of (a) ammonia
and (b) niotrogen dioxide.


    The quantitative response of a polypyrrole sensor to a sequence of
pulses of 0.1% nitrogen dioxide on the same timescale is shown at
Figure 5. Again the response (in this case a resistance decrease) is
sustained over eight pulses during a two hour period, but the
background resistance shows a progressive increase.
    Sensors could also be made for hydrogen sulphide, but in this case
there was evidence of an irreversible surface reaction possibly caused
by gas retention by the testing apparatus. However, tests in an open
environment yielded reversible results over a wide range of gas
concentrations from 100% down to 0.1% $H_2S$ in air. The response to a
series of pulses of 0.1% hydrogen sulphide in air in a closed
environment was a resistance decrease from which recovery was rather
slow so that, after a small number of cycles (Fig. 6) the response was
lost. The sign of this response is the opposite of what would be
expected for a p-type material behaving according to the tin dioxide
model (2). The mechanism of the response to hydrogen sulphide is not
yet understood.

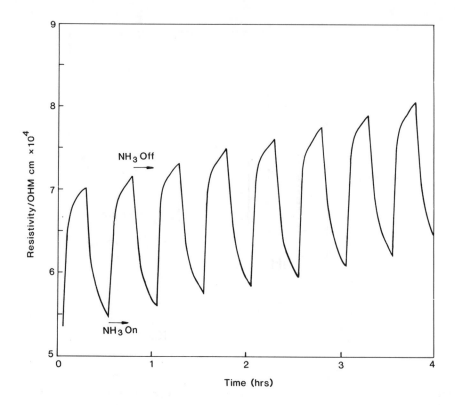

Figure 4.  AC (1 kHz) resistivity changes at room temperature for a conducting polypyrrole sensor exposed to eight 15 minute pulses of 0.1% ammonia in air over a period of 4 hours.

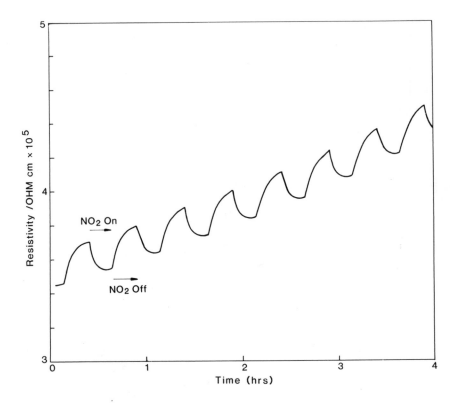

Figure 5. AC (1 kHz) resistivity changes at room temperature for a conducting polypyrrole sensor exposed to eight 15 minute pulses of 0.1% nitrogen dioxide in air over a period of 4 hours.

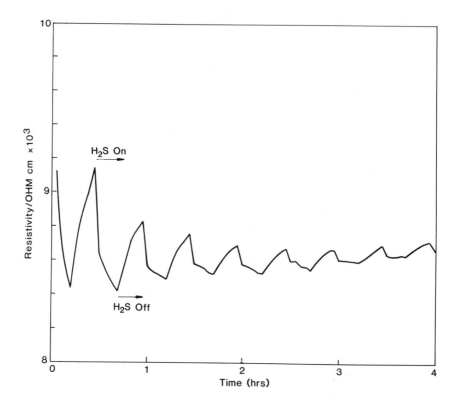

Figure 6. AC (1 kHz) resistivity changes at room temperature for a conducting polypyrrole sensor exposed to eight 15 minute pulses of 0.1% hydrogen sulphide in air over a period of 4 hours.

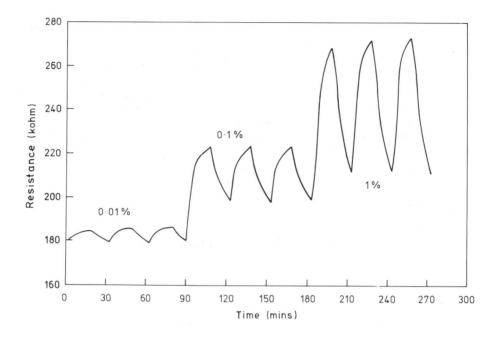

Figure 7. Room temperature AC (1 kHz) resistance changes for a conducting polypyrrole sensor exposed to 15 minute pulses of (a) 0.01%, (b) 0.1% and (c) 1% ammonia in air.

An important parameter to be considered during an assessment of the potential utility of new materials in gas sensing is the variation in response with gas concentration. Figure 7 shows the reproducibility of the polypyrrole sensor to pulses of ammonia at three different concentrations. There is clearly a useful response over a wide concentration range.

4. CONCLUSIONS

The results presented here confirm the sensitivity of polypyrrole to the presence of ammonia in air and show that the material is also sensitive, when prepared in the form of a thin film and post-treated, to other important toxic gases.

Gas sensitive devices based on a conventional interdigitated electrode array are straightforward to fabricate, robust and simple to use. Depending on the preparative conditions, devices may be made sensitive to a particular gas and insensitive to others although a detailed understanding of the factors affecting selectivity is not yet developed.

The responses of polypyrrole sensors to ammonia and to nitrogen dioxide appear to be reversible. The reversibility of the response to hydrogen sulphide appears to be system-dependent. There appears to be no interference by flammable gases such as hydrogen, methane and carbon monoxide or by carbon dioxide. The feasibility of using conducting polymers for the monitoring of toxic gases at room temperature has been clearly demonstrated although there remains a good deal of development work to be tackled before the mechanisms of response are fully understood.

ACKNOWLEDGEMENTS

The authors gratefully acknowledge financial support from the Department of Trade and Industry and experimental assistance from Mr. P. Taylor.

REFERENCES

1.  P.T. Moseley and B.C. Tofield, (1985), Mater. Sci. Techol. 5, 1, 505.
2.  J.F. McAleer, P.T. Moseley, J.O.W. Norris and D.E. Williams, (1986), J. Chem. Soc. Faraday Trans., in press.
3.  B. Bott and T.A. Jones, (1984), Sensors and Actuators, 5, 43-53.
4.  C. Nylander, M. Armgarth and I. Lundström, Proc. Int. Meeting on Chemical Sensors, Fukuoka, (1983), Ed. T. Seiyama, K. Fueki, J. Shiokawa and S. Suzuki, (Elsevier, Amsterdam, 1983), 203-207.
5.  J.J. Miasik, A. Hooper and B.C. Tofield, (1986), J. Chem. Soc. Faraday Trans 1, 82, 1117-1126.

REPORTS ON THE PANEL DISCUSSIONS

# 1 - PREPARATIVE METHODS AND PROPERTIES OF WELL ORIENTED ELECTRONICALLY CONDUCTING POLYMERS

H. Shirakawa (Discussion Leader)
University of Tsukuba, Sakura-Mura, Ibaraki 305, Japan

Kenneth J. Wynne (Reporter)
Office of Naval Research, Arlington, VA 22217-5000, U.S.A.

Abstract
    Orientation effects play an extremely important role in macro-molecular science. orientation can determine whether a polymeric material is a useless powder or a high strength fiber, a film easily split or tough, optically polarizing or isotropic. Orientation is also important in determining the properties of electronically conducting polymers. In this conference, considerable attention was given to methods of achieving orientation and resulting properties. After a brief introduction, the discussion of orientation is presented in the following order: (1) methods of achieving orientation, (2) methods for obtaining single crystals.

Chain molecules, including those which contain conjugated unsaturated structures, may be highly oriented due to the manner of synthesis or processing. It is not surprising that electronic motion is found dependent on macromolecular orientation, as high electronic conductivity in organic polymeric solids requires the presence of a conduction band formed by the overlap of pi-orbitais. Orientation in polymers is of great interest, as it may be used to improve polymer properties or to assess mechanisms of electronic transport phenomena.

Methods of Achieving Orientation. The most common ways of achieving orientation in semicrystalline polymers are spinning from a gel, melt or solution, stretching in the solid state, or solid state extrusion. Biaxial orientation is found in blown films or in films prepared by passing a melt over a conical die.

Unfortunately, few conducting polymers are processable by these conventional techniques which generally require the existence of an accessible glass transition or melting temperature. Ionic forces in the solid state tend to make partially oxidized, highly conducting polymers intractable.

One method of achieving orientation in a conducting polymer is the preparation of a prepolymer which is tractable. P.D. Towsand presented results on "Durham" polyacetylene (J. Feast) which is prepared by the thermal decomposition of a percursor polymer. A 20:1 ratio of $\ell$ to $\ell_0$ may be obtained for the polyacetylene prepared in this way.

L. Alcácer (ed.), Conducting Polymers, 201–203.

The resulting $(CH)_x$ exhibits a well defined fiber pattern and provides the opportunity of determining whether intrachain or interchain processes are most efficient for generating photoinduced carriers. Towsand's work on photoexcitation and charge transport in highly oriented Durham polyacetylene showed that there is a factor of 4 favoring interchain vs. intrachain motion.

H. Shirakawa gave an overview of methods for obtaining highly oriented polyacetylene. He described a new method using a nematic liquid crystalline host as a substrate for film growth. The films were grown in the presence of a magnetic field ($H_0$ = 10 Kgauss) to give well oriented fibrils. The conductivity parallel to the fiber axis was $1.2 \times 10^4$S cm$^{-1}$, while that perpendicular to this direction was $4.8 \times 10^3$ S cm$^{-1}$.

H. Naarman reported a new variation on the Zeigler-Natta route to polyacetylene. Naarman utilized a standard Ziegler-Natta catalyst in silicone oil at ambient temperature. This method gives a product which can be highly oriented (to 540%). The conductivity along the stretch direction is the highest reported thus far ($1.7 \times 10^4$ S cm$^{-1}$). Furthermore, a study of conductivity vs. time showed that iodine doped, highly oriented polyacetylene retained high conductivity in air for much longer periods of time than either unoriented or "Shirakawa" polyacetylene.

Other conducting polymeric materials exhibit anisotropic behavior. Conducting metal phtalocyanine/Kevlar[R] molecular/macromolecular blends (MMB's) have an anisotropic structure (Inabe, Marks, Wynne), but the conductivity appears isotropic or nearly so (peak conductivity is 5 S cm$^{-1}$). Fibers of doped MPcI$_x$/Kevlar[R] may be spun out of strong acid solvents reflecting a degree of processability. The solid consist of microcrystallites of MPcI$_x$ oriented along the fiber axis in an oriented crystalline Kevlar[R] matrix. At MPcI$_x$ loading levels of 40-50% by weight, where conductivity is nearly Maximum, a high modulus is retained.

Methods for Obtaining Polymer Single Crystals. R. Baughman pointed out that it is not possible to assess orientation definitively on the basis of a measurement such as conductivity, which is sensitive to effective conjugation length. He noted that it is best to combine measurement techniques, e.g., x-ray and spectroscopic methods. For the ultimate in detailed structural (x-ray) and spectroscopic (vibrational, optical) characterization single crystals required. H. Shirakawa noted that $(SN)_x$ and polydiacetylene were the only polymer single crystals with unsaturated electronic structures.

To set the problem of obtaining single crystal conducting polymers in perspective, R. Baughman presented a general outline of methods for obtaining three dimensional order in organic polymeric solids. Three approaches were mentioned in the discussion:

a) Simultaneous crystallization/synthesis (e.g., Wunderlich,s work on the polyoxymethylene).

(b) Solid State polymerization-polydiacetylenes (pionneered by G. Wegner).

(c) Matrix controlled polymerization - cyclohexadiyne in thiourea

Baughman commented that even if one had a single crystal starting material, partial oxidation would likely give rise to a less ordered structure. He noted that solid state reactions do not work in general to give an ordered structure. Problems which arise include reaction non-uniqueness and Van der Waals volume changes during reaction (a 25% volume change is typical in going from monomer to polymer and results in loss of template during synthesis).

In the context of this discussion, A. Epstein commented that one must always question the structure of a highly conducting material after doping relative to the structure of the starting polymer. There may be clustering (i.e., structural inhomogeneity) associated with dopant species (cations or anions). Generally, long range order is not observed. With the "new" structure after partial oxidation the question arises as to whether there will be a corresponding "new" magnetic behavior.

Electrochemical growth of charge transfer complexes was noted and A. MacDiarmid speculated that electrochemical growth of a conducting polymer might lead to a single crystal. The consensus was that this would be a most interesting and promising route which seems to have been overlooked.

Single crystals can give definitive physical information and oriented polymers may be useful for composite structures. However, B. Scrosatti pointed out that oriented materials generally have poorer kinetics with regard to electrode processes due to slower ion migration.

2 - PROSPECTIVES OF REALIZATION OF POLYMER ELECTROLYTES WITH AMORPHOUS
STRUCTURES AND CONSEQUENTLY HIGH CONDUCTIVITY AT ROOM TEMPERATURE

G.C. FARRINGTON (Discussion Leader)
University of Pennsilvania, Philadelphia, PA 19104-6272, U.S.A.

B. SCROSATTI (Reporter)
University of Rome, P. Aldo Moro, 00185, Italy

G. Farrington opened the panel discussion with some introductory remarks on the relevance of the amorphous phase of polymeric electrolytes
for assuring high conductivity and on the possible routes which may be
followed to lower the temperature range of stabilization of this phase.

M. Gauthier reported that chemical modifications have been performed
on PEO in order to lower Tg and thus to obtain a polymeric complex having
conductivity of the order of $10^{-4}(\Omega cm)^{-1}$ at room temperature. This
electrolyte has been used to realize a thin-film cell based on a $TiS_2$
cathode. This cell can be discharged and cycled at low rate (i.e. C/28,
about 5 $\mu A/cm^2$). Further improvements in the modification of the polyether chain are foreseen to obtain electrolytes which may allow the
realization of cells capable of being discharged at C/20 rate and at
temperatures down to $0\,^{\circ}C$.

With these electrolytes one can pile up various cells to develope
thin-film batteries having high voltage ( 50 V).

A discussion arose on the possible practical applications of the
room temperature, polymeric electrolyte batteries. At the present stage
of development of electrolyte modification and procedures, these applications seem to be limited to some special, low-rate microelectronic
devices. An increase of at last one order of magnitude in conductivity
of the electrolyte is considered to be necessary to broaden the field of
application.

Various directions to accomplish this goal were discussed. The most
promising appear to be cross-linking, addition of plasticizers and polyether chain modifications.

R. Spinder reported that $MEEP_4LiCF_3SO_3$ electrolytes (where MEEP stays
for $[P=N]$ , R = $O(CH_2-CH_2O)_x CH_3$ with O/Li = 16, have a conductivity at
$40\,^{\circ}C$ of about $5\times10^{-5}\ ^{-1}cm^{-1}$. The $Li^+$ transport number, however, in
these electrolytes is limited within 0.2 - 0.5.

B. Scrosatti suggested that low ionic conductivity at room temperature may be a problem for battery applications, but it may be useful for
other devices. In fact, the the large jump in conductivity (about 3 orders of magnitude) in a small temperature range (about $40\,^{\circ}C$) may be
conveniently exploited to develope thermal sensors.

The importance of a correct measurement of diffusion coefficients

L. Alcácer (ed.), Conducting Polymers, 205–206.
© 1987 by D. Reidel Publishing Company.

of ions, as well as of separation between electronic and ionic conductivity, in view of a proper characterization of the polymer electrolytes
and of a realistic prevision of their applications was stressed  by
various speakers in the audience.

## 3 - INTERFACES AND CONTACT PROBLEMS

B. SCROSATTI (Discussion Leader)
University of Rome, P. Aldo Moro 5, 00185 Rome, Italy

R. MAMMONE (Reporter)
E.T.D.L., Fort Monmouth, N.J. 07703, U.S.A.

At the beginning of this panel discussion, two subjects were introduced for open discussion. The first subject dealt with investigation
of the lithium/solid polymer electrolyte interface. Based on A.C. complex plane impedance measuments, passivating layers are believed to be
formed on the lithium when in contact with solid polymer electrolytes.
During charging, this passivating layer may be responsible for non-
-uniform lithium deposition which can ultimately lead to dendrite formation. However, since these cells are usually fabricated in quite thin
configurations in order to minimize IR drops across the solid polymer
electrolyte, the large effective surface areas and resulting low current
densities reduce the likelihood of dendrite formation. Several hundred
cell cycles are possible. The second subject dealt with investigation
of the interface between a polymer electrode and an organic liquid.
Based on a number of studies, these polymer electrodes undergo a spontaneous self-discharge process upon standing. An understanding of
factors contributing to loss of charge upon standing is crucial to the
further development of polymer electrodes.

The first subject on the lithium/solid polymer electrolyte interface
generated a discussion on the problems of employing A.C. complex plane
impedance as a tool to investigate surfaces. It was pointed out that
A.C. impedance data is relatively simple to acquire, but data interpretation is often very difficult. Some problems associated with the
techniques include (i) elimination of contributions from the counter
electrode and the solution, (ii) identification and use of a suitable,
completely reversible, reference electrode, and (iii) sensitivity to
impurities (e.g., water). It was suggested that electrode surfaces could
be more appropriately studied by spectroscopic techniques (e.g.,in-situ
FTIR, RAMAN).

The discussion then focused on problems associated with suitable
reference electrodes for A.C. impedance studies and potential usefulness
of spectroscopic techniques to study the lithium/solid polymer electrolyte interface. Since solid polymer electrolytes have lower conductivities, the location of reference electrodes may be critical. Furthermore,
since no convection occurs in these solid electrolytes, concentration
gradients can develop. It was suggested that since these electrolytes

*L. Alcácer (ed.), Conducting Polymers, 207–208.*
© *1987 by D. Reidel Publishing Company.*

have low vapor pressures, spectroscopic techniques (e.g., XPS) normally not applicable to liquid system might be more suitable.

The general conclusion was that for studying solid polymer electrolytes, electrochemical studies are best combined with traditional surface techniques.

The second subject which concerned the interface between a polymer electrode and an organic liquid, generated a discussion about inherent stability of oxidized (p-doped) and reduced (n-doped) forms of conducting polymers. It was mentioned that one must first look at chemical stability of the oxidized and reduced form of the conducting polymer in a vacuum. Oxidized polyacetylene, $(CH)^{+y}(A)^{-}_{y\ x}$, decomposes at different rates with different counter anions even in a vacuum and this decomposition rate may be expected to increase in liquid electrolytes. In vacuum, reduced polyacetylene, $(M)^{+}_{y}(CH)^{-y}_{\ x}$, is thermally stable with no chemical decomposition to 300°C in vacuum. In addition, reduced polyacetylene exhibits excellent stability in electrolytes such as 1M LiClO$_4$/THF. This excellent stability is probably due to formation of a passivating layer.

The general conclusion on this subject was that the chemical stability of conducting polymers must first be investigated in vacuum before further studies can be carried out in various electrolytes. If the polymer is stable in vacuum, other factors such as impurities, electrode morphology, formation of a passivating layer, chemical reaction with the electrolyte should be studied for their contribution to the conducting polymers stability or instability in the electrolyte.

4 - CYCLABILITY OF POLYMER ELECTROLYTE CELLS. POWER EFFICIENCY AND
    ENERGY DENSITY

J.O' M. BOCKRIS (Discussion Leader)
Texas A. & M. University, College Station, Texas 77843, U.S.A.

M. GAUTHIER (Reporter)
Hydro-Quebec, Varennes, Quebec, Canada JOL 2P0

   The panel discussion was actively animated by J. Bockris,
Discussion Leader, who stressed from the beginning the need to correctly
define the theoretical energy content of a battery (Wh/Kg or Wh/1) using
the mean voltage of the theoretical discharge plateau and including the
weight, or volume, of all chemical reactants involved in the electrode
processes. A factor of 1/4 to 1/6 is usually observed between practical
and theoretical energy content of current batteries because of dead
weight: packaging, collectors, electrolyte... and inefficient electrochemical processes. It was recalled that conducting polymer batteries,
like any other electrochemical generator, are inevitably subject to
electrochemical phenomena that tend to reduce their energy efficiency;
it is the case with IR drops or with activation and diffusion over-
potentials present in the electrodes, electrolyte and at the interfaces.
The members of the panel concluded to the necessity of defining the
practical energy content of conducting polymer batteries in association
with the power at which this energy is delivered, preferably in the form
of conventional Ragone Plots: (Wh/Kg vs W/Kg) or (Wh/1 vs W/1). Self-
-discharge and Cycle life were also identified as important characteristics that must be presented rigorously: it was suggested, for example,
that deep-discharge cycle results (over 60%) be reported as they are
much more demanding than superficial cycling.
   In this context the performance of electronically conducting
polymer batteries was very briefly commented by A. MacDiarmid and R.
Baughman. The state of the art of the Allied project was stated as a
55 Wh/Kg energy capacity at a typical discharge rate of C/10, no significant capacity decline over 250 cycles and a self-discharge less severe
than Nickel/Cadmium battery. Discussion on electronically conductive
polymer batteries was cut short, by the Discussion Leader as the Allied
Chemical scientist declined to elaborate on any further results for
confidentialty reasons.
   Thin film batteries made of ionically conductive polymers and
more conventional electrode materials, e.g. $TiS_2$, VOx and Lithium metal,
were discussed afterwards by G. Farrington, B. Scrosatti, J. Owen and
M. Gauthier. Despite the scarcity of published literature by institutions
active in this field, the development of new polymer electrolytes and the

*L. Alcácer (ed.), Conducting Polymers, 209–210.*
© *1987 by D. Reidel Publishing Company.*

performances of thin film batteries seems to progress actively.
          Small cells (a few square centimeters) based on
Li /PEO-LIX/$V_6O_{13}$ (or $TiS_2$) have already been tested over 400-500 deep-
-discharge cycles at close to 100°C, (Harwell and Hydro-Quebec/SNEA
works). More recently such cells have been scaled-up to a 10 Wh prismatic
cell, operated at 95°C over more than a hundred full discharge cycles,
without any loss of performance over smaller cells (1-a). A Ragone plot
deducted from this scaled-up cell is illustrated in Figure 1; realistic
expectations for future cells, including packaging, could be in excess
of 100 Wh/Kg and 100/Kg associated with cycling capabilities over 500
cycles (Harwell and H-Q/SNEA projections for warm batteries, 80-120°C).

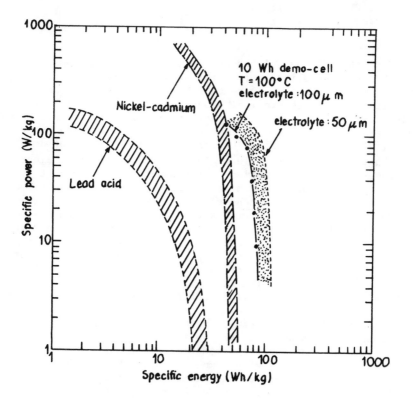

Fig.1 - Ragone plot corresponding to 10 Wh demonstration cell

References:

(1-a) 3rd International Meeting on Lithium Batteries, 27-30 May 1986,
      Kyoto, Japan, Extended Abstracts, Abstr. ST 11, p.238.

CONCLUSIONS

# 1 - CONCLUSION ON ELECTRONICALLY CONDUCTING POLYMERS

R. BAUGHMAN
Allied-Signal Inc., Morristown, N.J. 07960-1021, U.S.A.

As long as conducting polymers remained nonprocessible or thermally or environmentally unstable, there was little hope for commercial applications other than in specialty areas. However, major improvements to the properties of conducting polymers have been recently obtained. For example, it is now known that the presence of flexible substituents can provide solution or melt processibility without dramatically decreasing conductivities in the doped state. Also, a variety of conducting polymers is now known which are air stable and have reasonably high thermal stabilities.

However, further property improvements are required for many applications. Most noteworthy, the thermal stability of conducting in acceptor-doped polymers needs to be increased if such conducting polymers are to serve in many engineering applications. There are several examples of conducting polymers which do have outstanding thermal stability, such as alkali-metal doped polyacetylene and pyropolymers such as irregularly-reacted poly(perinaphtalene). However, this thermal stability was obtained at a cost - in the former case by complete loss of air stability and in the latter case by sacrifice in processibility. Mechanical properties also need improvement for many conducting polymers. The problem of poor mechanical properties is probably most easily dealt with, since low molecular weight is one cause and the absence of mechanical loss-mechanisms for rigid polymers is another. Improvements in synthetic methods can eliminate the former and suitable choice of substituents can eliminate the latter.

Processibility still remains a concern, since processibility appears to be much easier to obtain for undoped conducting polymers than for the same polymers in doped forms. If the polymer must be doped after fabrication of a shaped article, the problem of introducing the dopant uniformly is serieous, especially when it is required that the same dopant which easily diffuses into the article must eventually be thermally stable. What is needed here appears to be new methods for processing conducting polymers. One attractive possibility would be a process analogous to conventional RIM (reaction injection molding) in which polymer articles are directly fabricated from a mixture of monomer and dopant.

L. Alcácer (ed.), Conducting Polymers, 213–214.
© 1987 by D. Reidel Publishing Company.

Despite the absence of both air stability and high thermal stability
for many conducting polymers, a number of applications can be seriously
considered using these materials. For example, major progress has been
made in using n-doped conducting polymers as anodes in high-energy-
-density, rechargeable batteries. Here, the battery design insures stability by protecting against air exposure. On the other hand, one application has been described which is based on the instability of conducting polymers. This application (developed at Allied-Signal Corporation)
employs conducting polymers as ambient-responsive layers on antiheft
targets. Depending upon the choice of such polymer layers, the antiheft
target is transformed into a variety of different types of remotely-
-readible indicatior devices: time-temperature, humidity, radiation-
-dosage, mechanical-abuse, or chemical-release indicators for in-box
monitoring of products.

The ultimately obtainable properties of conducting polymers  are
clearly far superior to those which have been achieved. For example,
defect-related limitations on the effective conjugation length for polymers such as polyacetylene and poly(p-phenylene) dramatically decrease
electrical conductivity. Interchain hopping appears to limit conductivities in the chain-axis direction, as well as for orthogonal directions
for these polymers. Upon improvement in conjugation length, structural
order, and chain alignment, dramatic increases are expected in chain-
-direction conductivities, mechanical modulus, and ultimate strength.
For example, fully chain-oriented, high structural perfection Li-doped
polyacetylene should have an ultimate strength and modulus in the chain-
-axis direction which exceeds that of steel - while having a density
that is only about 1/5 that of steel.

# 2 - CONCLUSION ON IONICALLY CONDUCTING POLYMERS

M. GAUTHIER
Hydro-Quebec, Varennes, Quebec, Canada JOL 2PO

Ionically conductive polymers are starting to be recognized as a unique class of electrolytes of great technological potential for applications in many fields. They are presently successfully applied to rechargeable lithium batteries as illustrated by recent published results. The major advantages of film solid state batteries result from the apparent compatibility of polyether complexes with all chemical reactants of the battery. Mechanical strength associated with flexibility of the electrolyte allow good cycling properties. Good energy and power densities are projected and partially demonstrated for warm batteries; at room temperature low to moderate power appear possible.

Inherent properties of polymer electrolytes such as ease of elaboration with large surface/thickness ratio, absence of convection, specific mechanisms for ion solvatation, eventual control of the charge carriers: anionic or cationic transport... shall certainly result in a multitude of electrochemical devices beyond the sole aspect of power generation. A particularly promising area will probably be the exploration of possible applications resulting from coupling ionically conductive polymers with numerous electronically conductive materials such as polypyrroles.

Future developments of ionically conductive polymers could certainly be accelerated by more fundamental research on the exact physical mechanisms involved in the transport of ions in those semi-liquid electrolytes. Improving the cationic conductivity of polymer-salt complexes is still a priority, especially at, and lower than, room temperature; in this respect the development of new generations of solvating polymers has to be pursued.

Adaptation of usual electrochemical techniques (three electrode devices) to thin film polymer technology was identified as highly desirable in order to improve the understanding of polymer electrolyte cell behaviour. In addition, specific properties of polymer electrolytes, like low vapor pressure, should be exploited to adapt physical analytical techniques to observation of polymer electrlyte cell components.

In regard to battery applications of conducting polymers, it is obvious that more demonstration of scaled-up designs are needed to establish on a stronger basis the characteristics of the several systems

*L. Alcácer (ed.), Conducting Polymers, 215–216.*
© *1987 by D. Reidel Publishing Company.*

proposed and to evaluate their future and application domains. The need
for rigorous comparison tests, like the Ragone representation, is stress
ed by many participants.

The development of new polymers for room temperature applications
is still difficult to assess; however lower performances are expected
since cationic conductivity of polymer electrolytes compatible with
lithium battery environment still appear low: $10^{-6}$ to $10^{-4}$ (?) $S.cm^{-1}$.
On the basis of known conductivities at room temperature and practical
considerations on electrolyte thickness, projections of power capacity
in the range of 3-30 W/Kg for the next few years appear realistic (J.
Owen). Such values are partially confirmed by recent results presented
on small cells (1-b,c) and on a scalled-up 1 Wk prototype operated at
$25^{\circ}C$ (2); in addition, over 100 full discharge cycles are presented
without significant capacity decline and the self-discharge rate is
found inferior to 1% a year.

Available results on batteries made with ionically conducting polymers clearly establish the feasibility and compatibility of those
electrolytes in the chemical environment of a rechargeable lithium battery. Apprehension about the aggravation of dendrite occurence during
the charge cycle and the mechanical reliability of electrolyte films,
tens of microns up to one hundred microns thick, did not materialize
after scaling up the technology to relatively large surfaces, up to
5000 $cm^2$. On the other hand, more research is needed, in particular to
improve the transport properties of polymer electrolytes at room and
under room temperature. This presently restricts the battery discharge
rate to C/10 over C/100 range. More development and demonstration work
are also required to really determine the potential of multi-layered
plastic batteries and their future in large or miniaturized installa-
tions.

References:

(1-a)  3rd International Meeting on Lithium Batteries, 27-30 May 1986,
Kyoto, Japan, Extended Abstracts, Abstr.   ST 11, p.238.

(1-b)  idem, Abstr.   ST 12, p.242.

(1-c)  idem, Abstr.   ST  8, p.228.

(2)    15th International Power Sources Symposium, 8-11 Spe. 1986,
Brighton, U.K., Extended Abstracts, Poster Section.

# SUBJECT INDEX

Activation Energy, 48
Adsorption, 28
AgI, 96
Air Stability, 213
Aldissi and Polyacetylene, 65
3-Aminopropyl Triethoxysilane, 185
Ammonia, 190, 191, 192, 193, 194, 195, 197, 198
Amorphous Polymer Electrolytes, 205
Amphiphatic Molecules, 180
Aniline, Polimerization of, 105, 117
Anthracene, 1
Antisoliton, 127
Antitheft Targets, 214
Artificial Heart, 5
Artificial, Nerves, 5
Azobenzene, 179, 181, 185, 186, 187
$Ba(SCN)_2 \cdot 3H_2O \cdot (PEO)_{6.5}$, Preparation, 89
Band Bending, 13
Batteries, 1, 214, 216
    Solid State, 89
Benzidine, 144
Biosensores, 11
Biphenyl(Benzidine), 141
Bipolarons, 7, 130, 131, 135
Bis-Azo PC, 183
Boguslavski, 3
Bond Alternation, 126
4(4'-N.Phenylazo)Phenylbutiric Acid, 185
$Ca(SCN)_2 \cdot 4H_2O \cdot (PEO)_{6.5}$ Preparation, 89

Capacitors, 130
Carbon Black, 103
Charge Injection, 2
Charge Transfer Complex, 48
Charge-Dipole, 61
Chemical Stability, 208
Chemisorption, 179, 180
Chlorpromazine Hydrochloride, 49, 58
$CO_2$, 5
Composite Electrodes, 103
Conjugation Length, 214
Corrosion, 29
Creation Energy, 128
$CrO_3$, 145
Crosslink, 153, 205
Cyclability of Polymer Electrolyte Cells, 209
Cycle Lyfe, 209
Decarboxylative Condensation, 142
Dendrite, 207, 216
Devices Based on Biological Macro-molecules, 179
    Based on Organic Molecules, 179
    Electrochemical, 205, 215
4,4-Diamino-E-Stilbene, 141, 144
2,7-Diaminofluorene, (DAF), 144
1,6-Diaminopyrene, 144
Diaz, 4
Diffusion Coefficients, 205
2,5-Dihydroxy-1,4-Dihydro-terephthalic Acid (DHDHTA), 143

Diisocyanobenzene, 25, 173, 174
Dimethylsiloxane, 151
Diodes, 23
Dipalmitoylphosphatidylcholine, 183
Dipole-Induced Dipole, 61,
Discharge Cycles, 216
DMF, 144, 145
Domain Wall, 128
Dopants, 6
Doping, Bronstead Acid, 141
    Electrochemical, of
    |PCFE(PYZ)|N, 177
Double Layer, 1, 13
DSC, 89, 91, 96, 99, 151, 152, 175
Durham and Polyacetylene, 77, 201
DYE Sensitisation Processes, 180
Elastic Energy, 126
Electrocatalytic Properties, 162
Electrochemistry, 1
Electrode, Catalitic, 161
    Insersion, 103
Electrodes, 1
Electron Bombardment, 181
Electron Transfer, 47
Electronic Conductivity, 103
Electronic Energy, 126
Emeraldine, 142
    Conductivity of, 105
    Electrochemical Studies of,
    105, 115, 119
Emeraldine Base, 105, 132, 133, 145
    Capacitance Effects, 119
    Equilibration With Aqueous
    HCL, 113, 118
    Synthesis of, 105, 108, 110
    TiCl$_3$ Reduction, 115,
    Analytical Data, 105, 109
    118
Emeraldine Hydrochloride, 105,
    110,112
    Pauli Susceptibility, 118
    Polaronic Conduction Band, 118
    Synthesis of, 105
Emeraldine Salt, 132, 133, 137
Emeraldine Sats, Traces of Water In,
    118
Energy Density, 209, 215
Energy Gap, 124, 125, 130
Enzymes, 185
Fluorene, 141
Fluorescence Energy Transfer, 180

Graft Polymerisation, 185
Honzl and Moore, 143
Host Lattice, 48
Hybrid Films, 181
Hydrogen Bonding Interactions, 61
Hydrogen Sulphide, 193, 196
Impedance Measurements, 28, 96, 151,
    152, 155, 207
Interface and Contact Problems, 207
Interface, Electrode-Solution, 1
Iodine Complexes, 52
Ion Bombardment, 181
Ionene-Iodine Complex, 58
Ionene-TCNQ Complexes, 58
Ionenes, 48, 58
Ionic Conductivity Measurements, 89,
    92, 95
IR Drops, 207, 209
IR Spectroscopy, 173, 175
Iron Phthalocyanines, 48, 50
K$_2$CO$_3$, 154
KAg$_4$I$_5$, Conductivity of, 95
Kallmann and Pope, 1
KI, 96
KSO$_3$CF$_3$, 154
Langmuir-Blodgett Film, 179, 180
Leuco Emeraldine, 105, 106, 133, 142
LiCF$_3$SO$_3$.(PEO)$_9$, Conductivity, 89,
    90
Linear Combination of Atomic
    Orbitals, 122
Liquid Crystalline Host, 202
LiSO$_3$CF$_3$, 152, 154
London Dispersion, 61
Lythium Battery, 216
Magnesium Mesophenyltetrabenzo-
    porphyrins, 49
Magnetic Susceptibility, 127, 133,
    134
Mechanical Modulus, 214
Mehl and hale, 2
Memory Devices, 23
Menschutkin Reaction, 49
Mesophenyltetrabenzporphyrins, 52
Metallo-Mesophenyltetrabenporphyrins,
    51
Metallocenes, 61
Metallomacrocycles, 173
Metalloporphyrin, 48
Methylsiloxane, 151
Methylthiophene, 161

MgCl$_2$-PEO Electrolytes, 89, 90
Micelles, 48
Michael Rice, 127
Micro-Electrodes, 25
Mobility, Segmental, 61
Molecular Electronics, 1, 179, 181
Molecular Engineering, 180
Mossbauer, 173, 174, 175
Mott-Schottky Plots, 28
Naphthalocyanine, 173, 175
NaSO$_3$CF$_3$, 154
Nickel/Cadmium Battery, 209
Nigraniline, 142
4-Nitrobenzil Triphenylphosphonium
     Bromide, 186
Nitrogen Dioxide, 192, 193, 198
NMR, 99, 151, 152, 153, 154, 173
Nucleopore, 26
Octaaniline, 142
Octamer, 142
Ohmic Contacts, 9
Order Parameter, 128
Oriented Electronically Conducting
     Polymers, 201
Overpotencials, 209
Oxigen-Silicon Linkages, 180
Palmityol Azo PC, 183
Passivating Layer, 207, 208
PbBr$_2$-PEO Electrolytes, 90, 91, 92
PbI$_2$ Electrolytes, 90, 91
PbO$_2$, 145
Peierls GAP, 124
Peierls Semiconductor, 131
Pendant Groups, 48
Pental Dimer of 10-(2-Methylamino-
     propyl)phenothiazine, 49
PEO, See also polyethylene Oxide,
     99
PEO$_8$LiClO$_4$, 89
Percolation Theory, 101
Pernigraniline, 106
Ph-Dependent Properties, 141, 144
Phase Segregation, 137
Phenothiazine Sulfides, 49, 51
Phenothiazines, 48, 51, 58
Phenyl-Capped Octaaniline (COA),
     142, 145
Phenylene, 142
Phosphatidylcholine, 183
Phosphazene, 151

Phosphonium Bromide, 186
Photochromic, 181
Photoelectrochemistry, 1
Photolithografic Methods, 181
Phthalocyanines, 173, 190, 202
Physisorption, 179
Plasticizers, 205
Polaron, 7, 129, 130, 131, 132,
     136, 137
Polyacetylene, 2, 37, 77, 79, 80,
     121, 127, 131, 132, 134,
     161, 201-203, 208
   Conductivity of Stretched, 70,
     79, 201, 202
   Crystalline Form, 77, 81, 201,
     202
   Crystalline Form, Photo-
     excitation, 81, 201
   Distributions of Conjugate
     segments, 38
   Doping, 70
   "Durham", 201
   Electron-Phonon Interactions,
     44
   EPR, 42
   Highly Oriented, 77, 201
   Hot Luminescence, 38
   I.R. Spectra, 72
   In Oriented Crystal Matrix, 65
   In Situ Raman Experiments, 38
   Morphology, 38
   Peierls Distortion, 38
   Raman, 37, 80
   Resonance Raman Scattering, 38
   S.E.M. of Oriented, 74
   Thermal Annealing, 43
   Transparent Films, 72
Polyaniline, 21, 38, 44, 105
     118, 121, 141, 181, 185
   Structure Determination, 142
Polybithiophenes, 149
Poly(Ethylene Glycol), 151
Poly(Ethylene Glycol Methyl Ether),
     151
Poly(Ethylene Oxide), Alkali Metal
     Complexes, 95
   Complexes, 95, 205
   Silver Halide Complexes, 95
   With Divalent Cations, 89

Poly(1,6-Heptadiyne), 132
Polymerisation, Electrochemical, 190, 192
Poly(Methylhydroxisiloxane), (PMHX), 151
Polyparaphenylene 44, 132, 214
Poly(Paraphenyleneamine), 106
Poly(Paraphenyleneamineimines), 105
Poly(Paraphenyleneimine), 106
Polyphthalocyanines, 173
    Conductivity of, 177
Poly(Propylene Oxide), 89
Polypyrrole, 4, 132, 149, 189, 190, 191, 192
    Sensor, 194
Polysemiquinone, 118
Polyterthiophenes, 149
Polythiophenes, 4, 132, 149, 163
Polyvinylpyridine, 49
Porosity, 9
Porphyrins, Stacked, 61
Power Density, 215
Power Efficiency, 209
Processibility, 213
Prosthetics, 2
Protonation, 134
Pyrazine, 173, 174
Pyrene, 141
Pyropolymers, 213
Quinone-Hidroquinone, 17
Quinoneimine, 142
Ragone Representation, 216
$RbAg_4I_5$, 95, 96
RbI, 96
Reaction Injection Molding, 213
Rectifier, Electrochemical, 23
Self-Discharge, 207, 209, 216
Semiconducting Polymers, 121
Sensor, for Ammonia, 189
    for Flammable Vapours, 190
    for Hydrogen Sulphide, 189
    for Nitrogen Dioxide, 189
    for Oxigen in Automobile
      Exhausts, 190
    for Oxigen, 190
    for Toxic Gases (CO, $H_2S$), 190
    of Polypyrrole Electrochemical
      Polymerised, 192
    Catalytic(Pellistor), 190
    Galvanic Solid Electrolyte, 190

Semiconductor, 190
    Tin Dioxide, 189
Sensors, 189, 192, 205
Serge Brazovskii, 127
Shirakawa and Polyacetylene, 19, 65, 201
Silicon, 127
Siloxane, 152, 153, 154
Siloxane Polymer Electrolyte, 151
Silver Paint Electrodes, 96
$SnCl_2$, 154
Solar Driven Aircraft, 6
Solid State Polymerization, 202
Soliton, 128, 129, 130, 136
    Charged, 83
Space (And SDI) Applications, 5
Stabilization Energies, 61
Structural Order, 214
SU Schrieffer and Heeger, 127
Surface Area, 9
Szent-Gyorgy, 2
Teflon, 3
Tetrabenzoporphyrine, 173
Tetrafluoroborate, 191
Tetramethyldiisocyanobenzene, 174
Tetrathiafulvalene-TCNQ, 61
S-Tetrazine, 173, 175
Thermal Stabilities, 213
Thermogravimetric Method, 90, 173, 175
Thin Film Technology, 179, 205
Thin-Film Batteries, 205
Tight Binding Approximation, 122
$TiS_2$, 209
TQI (Phenyl-Tetraquinone-Imine), 145
Trans-Polyacetylene, 123
Transfer Integral, 123
Transistors, 23
Transport Numbers Measurements, 92, 205
Trifluoromethanesulfonate, 151
Tunelling, 137, 180
$V_6O_{13}$, 103, 209
Vogel-Tamman-Fulcher Equation, 151, 157
Voltametry, Cyclic, 161
Zinc Mesophenyltetrabenzoporphyrins, 49
Zn Octoate, 152
$ZnCl_2$, 154